CULTIVATED PLANTS
OF THE FARM

CULTIVATED PLANTS
OF THE FARM

By

G. D. H. BELL, B.Sc., Ph.D.

Fellow of Selwyn College, Cambridge
University Lecturer in Agricultural Botany
Director of the Plant Breeding Institute
School of Agriculture, Cambridge

CAMBRIDGE
AT THE UNIVERSITY PRESS
1948

CAMBRIDGE UNIVERSITY PRESS
Cambridge, New York, Melbourne, Madrid, Cape Town,
Singapore, São Paulo, Delhi, Tokyo, Mexico City

Cambridge University Press
The Edinburgh Building, Cambridge CB2 8RU, UK

Published in the United States of America by Cambridge University Press, New York

www.cambridge.org
Information on this title: www.cambridge.org/9781107662797

First published 1948
First paperback edition 2011

A catalogue record for this publication is available from the British Library

ISBN 978-1-107-66279-7 Paperback

CONTENTS

PREFACE

The purpose of this book is to provide a short and non-technical account of the cultivated farm plants in Britain, and to discuss their importance to British agriculture. No attempt has been made to give detailed botanical descriptions of the plants, and the accounts are based on the most general botanical, agricultural and economic characters which make the plants valuable and worth cultivating. The important crops have been grouped together in their families, genera and species so that their botanical relationships and similarities may be appreciated, and the botanical names are included to clarify the rather confusingly large number of forms which characterises many of the species and genera.

Agriculture is still the world's largest and most important industry, and providing sufficient food for many hundreds of millions of people is agriculture's problem and task. Plant growth is the basis of agricultural production, and the development of cultivated food plants has made possible the intensive crop production that feeds man and his domestic animals. From time to time wars, crop failures and other world disasters remind us only too clearly how comparative plenty can give way rapidly to the misery and discomfort of want. But even in 'normal' times there are many millions of people who are not properly fed and clothed, and the future expansion that will be necessary in agricultural production will depend largely on an ever greater and more efficient output of plant products.

There are, however, limits to the expansion in agricultural production in spite of the wider application of more efficient methods. There are no new large areas in the world waiting to be taken into cultivation where a high standard of productivity can be expected, and a large proportion of the uncultivated land of the world is not capable of supporting the economic production of the known valuable food crops. Cultivated plants, crop husbandry and agriculture in general must be adapted to the natural conditions if there is to be a stable and permanent food production, and systems of farming grow up as a result of the adaptation of crops and stock to climate, soil, economic conditions and social customs. To understand fully the agricultural and economic significance of the cultivated plants that characterise the crop husbandry of a country, it is necessary to appreciate the conditions and circumstances that are responsible for the choice of the plants and the development of the particular type of husbandry.

The practice of cultivating plants has had far-reaching and important effects on the growth and stabilisation of civilisation, and as long as there is an increasing world population and an expanding economy, continuous improvement and development in crop plant production will be needed. The demand for food is constantly rising as numbers increase and standards of nutrition improve, but these trends can continue only as long as the production of food is intensified, and the quality reaches a higher level. The most promising methods of obtaining this increased quantity and improved quality are by better methods of production and improved varieties of the crop plants. In recent years, plant breeding has made some important contributions to agricultural production by providing the grower with new varieties and strains which are higher yielding, of better quality, more disease resistant, or which in some ways are better adapted to particular conditions, or more suitable for cultivation. Some of the most important problems facing improved crop production can only be solved by plant breeders, whose task it is to see that the farmer is provided with the most economically productive varieties, and the consumer receives a sufficient quantity of the best quality article.

It is, then, on the agricultural variety and strain that any study of the cultivated plants of the farm must finally focus attention. These varieties and strains may constitute a huge range of botanical forms in any one crop, as they do in the oldest crop plants such as wheat and barley, but the forms cultivated economically in any one country are usually a mere fraction of the total. The progressive development of crop husbandry depends very largely on the success with which improved varieties can be made available, and the extent to which the farmers take full advantage of these improvements. The modern agriculturalist is undoubtedly very conscious of the importance of the variety question, and there is, perhaps, no other technical problem of crop husbandry of such wide interest and importance as crop plant improvement.

No attempt has been made in this book to name and describe all the varieties and strains of each of the crop plants discussed, nor have the names of individual breeders, plant breeding institutes, or seeds firms responsible for the breeding of new varieties been given. The inclusion of such a large amount of detailed information is not suited to a book of this type, and all that has been attempted is an account of the more important types grown with a discussion of the significant accomplishments in plant breeding which have influenced the agricultural and economic development of crop husbandry. Lists of varieties and strains of crop plants are apt very quickly to become out-of-date, and the relative importance of individual named sorts may change in a matter of a few years.

The varieties and strains named, therefore, are not intended to include all the forms cultivated in Great Britain. This is particularly true of the strains of herbage plants, the numbers and diversity of which are becoming a serious problem not only in any discussion of crop plant characteristics peculiar to the agriculture of a country, but also in the more practical considerations of seed production, advisory work and crop testing. The great increase in the number of recognised strains, even within one species of certain herbage plants, is a matter which is going to require the most serious attention on the part of those responsible for marketing agricultural seeds, and also of agricultural officials whose duty it is to see that the farmer is using the best strains for particular conditions and purposes.

ACKNOWLEDGEMENTS

Much of the subject matter in a work of this kind cannot, of course, be original. In this particular book, in addition to the elementary scientific facts, which are common knowledge to specialists in various fields of study, and occur in books and scientific papers generally, information has been drawn from other sources, particularly official publications relating to crop statistics.

The photographs form an important and integral part of the book. These have been supplied by the courtesy of a number of people, representing various periodicals, publishing agencies, and corporations. Thanks and acknowledgements are gratefully offered to the following:

The Farmer and Stockbreeder: Plates 1, 2, 3, 4, 7, 8, 9, 10, 11, 13, 14, 15, 16, 17, 18, 22, 23, 25, 26.
Sport and Country: Plates 12, 19, 27, 28, 29, 30, 31, 32, 33, 34, 35, 36.
Sport and General: Plate 24.
Ministry of Information, Crown Copyright: Plate 20.
British Sugar Corporation: Plate 21.

Mr V. Chapman of the Plant Breeding Institute took the photographs for Plates 5 and 6.

This book is the outcome of an invitation from the Cambridge University Press to write an account of the farm plants of this country. Although as it stands it is not the book originally envisaged by the Syndics of the Press, it was their invitation which provided me with the opportunity to write it. I have tried to make it the kind of book which I have felt for some time should be of use to agricultural students at universities and colleges, as well as a book of general interest both to students of botany, and to a wider circle of readers who are generally attracted by agriculture and the story of our cultivated farm plants.

G. D. H. BELL

SCHOOL OF AGRICULTURE
Cambridge

ILLUSTRATIONS

Chapter I

THE BEGINNING AND DEVELOPMENT OF THE CULTIVATION OF CROP PLANTS

The domestication of plants and animals were critical stages in the history of man, and marked the beginning of a more secure way of living than that offered by the precarious existence of the hunter, the fisherman and the collector of the edible parts of wild plants. Tilling the soil, and growing plants in prepared ground where they could be freed from the struggle with other plants and protected from wild animals, offered an obvious way of providing a larger and more assured supply of vegetable foods, and, given suitable conditions, this method of food production presented new possibilities for life in settled communities. The cultivation of plants was necessary for the development of civilisations capable of steady advancement because of the much greater security that it provided, and the increased scope for other activities that resulted from the release from the incessant search for food. Not only was it possible to supply food and other necessary agricultural products on a more intensive scale, but there was a considerable reduction in the dangers of seasonal food shortages and famines. With improved methods of cultivation, and the selection of better varieties of plants, greater opportunities presented themselves for the development of other occupations, social activities and recreations, all of which are necessary for the growth of civilised life as opposed to bare subsistence.

It is not known when, or under what circumstances, the most ancient of the known cultivated plants were first domesticated, and the practice was obviously introduced at different times in the various early human settlements. The first attempts probably consisted of scattering the seeds of native wild plants in odd patches of bare soil near the dwellings, and little or no attempt was made to cultivate the ground or tend the plants. There must have been a long period in man's history from this haphazard method of collecting seeds and sowing them in the most

I

primitive fashion, to the time when the seed was kept and sown as a regular practice, and the land was properly cultivated by implements. Under domestication the first selection of new forms better suited to the conditions and to cultivation took place, and so started the long history of these plants which have persisted from prehistoric times until to-day as essential features of civilised life.

Many attempts have been made to trace the earliest history of the oldest cultivated plants, but the most ancient records merely show that the staple food crops, such as wheat and barley, which were then in cultivation, must have had many thousands of years of previous history under domestication. The legends that were current in the ancient civilisations of Mesopotamia, Africa, India and Asia show that the origin of the staple crop plants had already been lost when these civilisations were flourishing, and the common belief was that these plants had been bestowed as a gift from the gods. Even the earliest archaeological evidence can do little more than demonstrate the importance of the major crops, not only as a means of feeding the people, but also in the internal economy, art and religion of the civilisations. In Mesopotamia, barley was used as a bread corn and for making fermented drinks, but it was also a current means of paying labourer's wages, and the by-products of fermentation had several valued uses. Both wheat and barley were used as the inspiration for many kinds of artistic design, and there is no doubt that the early civilisations lived in very intimate association with the crops which were recognised as being the source of the people's material welfare.

The closeness of this association between the ancient civilisations and the important plants that they cultivated was, of course, accentuated by the very limited communications that existed between the established centres, and the virtual self-sufficiency that existed in each. This resulted in each separate civilisation being dependent on a limited number of crop plants, and in some cases one major crop, such as a particular species of wheat or barley, became the dominating influence. This is particularly true of the cereals, which were the first important stabilising influence for man, and which gave the grain growers the opportunity of making the first material advances in civilisation.

The extreme age of the oldest cultivated plants makes it very difficult to trace their wild ancestors, or 'progenitors', although there are 'primitive' forms still in existence in some cases which suggest how the cultivated forms originated. The more recently introduced cereals like oats and rye, which were probably not seriously cultivated before the Christian era, show a much clearer relationship with wild and primitive forms, and their histories can be visualised more easily. There are in oats, for example, small 'weed' kinds which are found only on un-cultivated soils; they are essentially wild in all their characters and have special mechanisms for seed dispersal and burial in the hard, uncultivated ground. But there are also weed oats, which grow principally on cultivated land and have larger and less weed-like grain, but they still retain the powers of seed dispersal and have a great capacity for mimicry according to the crop in which they grow. The known history of cultivated oats suggests that the plant was originally taken into cultivation under conditions where the weed became such a nuisance that it usurped the place of the crop, and cultivated types which did not shed their grain of their own accord were automatically selected.

Although the oat story is in some ways unique, it illustrates two important aspects of the domestication of plants—first, wild plants are the original source of cultivated plants; and secondly, when a plant is taken into cultivation it will be subject to changes brought about by its management as a crop plant. There are many cases of more recently domesticated crop plants where there is a fairly easily recognisable relationship between the cultivated plant and forms still growing wild. There are wild forms of potato growing in south and central America which are obviously related to the domestic potato grown in Europe; the wild sea beet occurs in many forms around the coasts of Europe, and is undoubtedly a blood relation of culti-vated sugar beet; while the carrot and the parsnip each has a wild counterpart belonging to the same species growing in this and other countries. But the process of domestication is most obvious in the herbage grasses where selection from wild forms is a very recent development and is being exploited at the present time for the production of forms suitable for 'cultivated' grassland.

The varying lengths of time for which the different crop plants have been cultivated, and the changes which have resulted from domestication, really decide whether it is possible to trace their botanical and economic histories. Some of the older cultivated plants very clearly show geographical centres where they occur in their greatest botanical profusion as the result of long periods of natural and artificial selection acting on the new forms which arise as the result of mutation and hybridisation. Growing plants in large numbers as field crops gives a better chance for hybridisation to take place than when plants occur singly or in small groups, and if the conditions are very variable, the numbers of forms will be correspondingly large. In this way, where agriculture has been practised for a long time in restricted areas that show a wide range of climatic and soil conditions, centres of botanical variation or diversity of the crop plants occur.

Such centres of diversity have been found, and they all occur in latitudes near the equator where the countryside is mountainous, but with valleys and mountain slopes which can be cultivated. There are more of these centres in the Old World than the New World, and they coincide with some of the most ancient centres of settled agriculture. Abyssinia, the Mediterranean region, Persia, Afghanistan, India and Burma, Malaya and Java, Siam, and parts of China are characteristic centres that between them can claim the greatest diversity of wheat, barley, oats, rye, rice and most of the temperate root crops, vegetables and fruits, as well as the important fibre plants with the exception of some cottons. On the American continent, Peru, Chile, Brazil and Paraguay, and Mexico appear as the centres of maize, potatoes, tobacco, Para rubber and cocoa; while a small area of the United States can lay claim to the sunflower and the Jerusalem artichoke.

These centres are interesting in showing the probable areas in which some of the oldest cultivated plants originated, although it is not always easy to localise and define clearly the actual geographical limit. They demonstrate also how dependent were the old civilisations on particular crops, such as the macaroni wheats in the Mediterranean region and the bread wheats in south-west Asia, while the more recent development of new centres of civilisation can be followed by the origin of new centres of botanical diversity of the crops. It has only been possible for man to colonise the vast new areas of the world during the more recent period of his history, because he has been able to select crop plants which can be grown successfully under the conditions which he has chosen for settlement.

The enormous increases in population which have accompanied the new areas of colonisation by man, as well as the ever increasing pressure of higher population densities in the older centres of civilisation, have resulted in a constantly expanding demand for more agricultural products. These new demands have been met by exploiting more intensively the available plant growth, and by increasing the output of the important crop plants. Something like four-fifths of the total world production of agricultural commodities at the present time are required for human consumption, and just over one-half is eaten as plant produce. But many countries find it impossible to grow sufficient food within their own territory, and a large international trade has developed in plant products which compose about three-quarters of all agricultural commodities traded in this way. In addition to food, these plant products include commodities such as timber, fibres, raw materials for beverages, and other necessities and luxuries that are important in the economy of many countries.

In spite of the vast increases in population and the economic and social developments that have taken place since the times when plants were first taken into cultivation, the cereals are still pre-eminent as providers of human food. Over four-fifths of the world's population of some two thousand two hundred million people relies on wheat and rice for its principal means of sustenance, while a tremendous amount of human food is obtained from rye, maize, millets and sorghums. These starchy foods, with that obtained from important tuber-bearing plants like the potato, are the cheapest source of energy available, and they contribute by far the greatest part of human diets. The production of this tremendous amount of food has been possible only by a great expansion of the area under these important staple crops, and the acreages now reached are of such proportions that it is only possible to estimate approximately their true extent. It can be said, however, that something like 400 million acres of wheat, 190 million acres of rice, 90 million acres of rye and 50 million acres of potatoes are grown each year, and even this is not sufficient to ensure that everyone is properly fed.

Although the production of food for human consumption is still the most important function of crop cultivation, it is not the dominant feature of the agriculture of all countries. Most countries import plant products of one form or another, and there is considerable trade in food for human beings and livestock. The conversion of plant products is, of course, an extremely valuable method of producing human food of special dietetic value, and meat, dairy produce and eggs have their own virtues. Economically speaking, the balance between crop and stock production in any agricultural system is one of its most important features, while the relative amounts of plant and animal foods that are consumed by the human population is a reflection of the standard of living. In some countries, as in the rice-growing areas of Asia, by far the greater proportion of the agricultural production is as plant produce, while nine-tenths of the calories in the people's diet is of cereal origin. The standard of living under these conditions is extremely low, and may be contrasted with New Zealand which produces and consumes a higher proportion of animal products than any other country, and contributes to the high standard of living of other countries by extensive exportation, as well as maintaining a high standard at home.

Not only have different kinds of agricultural production been developed in various parts of the world, according to the types of crops grown and the balance between crops and stock, but the position that agriculture occupies in the national economy varies from country to country. When a country has no other industries or natural resources than agriculture, and when the conditions support only a poor or restricted crop husbandry, there is little to offer the people beyond a peasant standard of living, dependent entirely on the products of locally grown crops. On the other hand, with good natural conditions and an enter-prising people, intensive crop and stock production can be developed in countries which are almost entirely dependent on agriculture for their incomes, so that exportable surpluses are available. In this way, particularly when the export products have a high cash value, a high standard of living can be maintained.

Many countries have important industries besides agriculture, and they may develop larger populations than they can support at a satisfactory standard of diet from home-produced food. In such cases, as much food is produced at home as the conditions allow and it is economic to do so, while the balance of human and livestock food is imported. These countries, and particularly over-populated and over-industrialised countries like Great Britain, cannot exist on the products of their own agriculture, and they are only able to support their large populations by drawing on the food production of other countries. Great Britain is an extreme example of this type of national economy, in which the money obtained from other industries and re-sources is used to buy food of all kinds from many parts of the world.

There have, therefore, been great changes since the time when crop cultivation consisted of the cultivation of native plants only, and the local population depended on the plants growing at their doorsteps. To-day the important staple food crops of the world have been introduced to all those areas where they can be grown successfully, and most of the important crop-growing regions rely on plants which are not natives. In addition to this direct introduction of plants for cultivation, there is a consider-able dependence in some countries on imported food to supple-ment the home production of the same crop, whether this is used for human or animal food. For countries that can afford the luxury, there is also the opportunity of importing food and other plant products that cannot be grown at home because of unsuitable climatic conditions. Such imports may be entirely luxuries, or else they may be important in helping to maintain intensive livestock industries which have assumed an essential place in the importing country's agriculture. In this way the world supply of plant products, and particularly that from cultivated plants, may be used far from the centres of intensive production, and the extensive distribution of these products is of great importance in maintaining the complicated structure of modern civilisation.

Chapter II

CROP PLANTS IN RELATION TO THE GROWING AND ECONOMIC CONDITIONS

The characteristic kinds of natural vegetation that are found under different growing conditions would indicate immediately that plants show individual requirements with regard to these growing conditions, and that particular forms of vegetation and associations of species are found under certain natural environments. When left undisturbed, vegetation is largely a reflection of climate, topography and soil, but as the climate has a marked effect on the soil that is developed in any area, it is largely the climatic influence that decides plant distribution and the structure of the natural vegetation. It is, however, very common for the naturally controlling influence of climate, and the more restricted action of the topography and the type of soil, to be subject to interference by the activities of man and animals. These activities, from the point of view of the vegetation, are largely destructive, and usually prevent the fullest development of the plant growth that the natural conditions allow. In the case of man, the most widespread destruction of natural vegetation has occurred in connection with his agricultural activities, particularly by the clearing of the land for the cultivation of the soil. When land is prepared in this way, it is possible to grow plants economically which would not survive if left to compete with the natural vegetation.

No agriculture can persist permanently unless it is adjusted to the natural conditions which set definite limits to the systems of farming which may be practised. When these limits are exceeded, agriculture may degenerate into exploitation, which sooner or later results in an uneconomic level of production, and may ultimately lead to devastation of the natural resources. The danger of developing types of farming which are unsuited to the environment is seen from the results in various parts of the world of maladjusted agriculture. The destruction of the natural vegetation, and the exposure of the soil to conditions and agri-

9

cultural practices which prevent the maintenance of the soil structure, have destroyed its capacity to carry plant growth. There are examples of this in parts of North America, South Africa, New Zealand and the tropics where the immediate causes may be different, but the general principles and results are the same. As a result of the agricultural exploitation, the soil becomes so eroded that it is not only incapable of supporting an economic agriculture, but there is great difficulty in re-establishing any stable form of vegetation.

Temperature, rainfall, sunlight and wind are the principal climatic factors that decide the distribution and types of natural vegetation, as well as determining the kinds of crops that can be grown successfully. Different kinds of plants have special requirements with regard to heat, light and moisture, and the actual conditions to which they are exposed are the result of the complicated action and interaction of the climatic factors. Not only does every plant species grow best within particular ranges of heat, light and moisture, but there are also special demands in relation to temperature and light exposure for sexual development, reproduction and the successful completion of the life cycle. The economic cultivation of plants requires more from the climate than merely its suitability for growth, and it is only when the natural conditions are particularly adapted to that type of growth and development which give the plants their agricultural and economic value that successful husbandry is possible. It is the climate that has as yet prevented the introduction into this country of the soya bean as a farm crop: the plant will grow, but the temperature and light condition are not suited to anything like economic yields of seed. Similar considerations limit the introduction into some countries of the potato, which only produces its highest yields of tubers when the plants are exposed to particular lengths of daylight during their growing season.

The complicated action of climate on plant growth is obvious when the seasonal variations in temperature, rainfall, light and wind are considered. The average temperature at different times of the year, the maxima and minima at different seasons, and even the difference between day and night temperatures are some of the more easily appreciated heat relationships that affect plant growth. Similarly, the amount and seasonal distribution of rainfall, the relative humidity of the atmosphere, the intensity of light and the length of day at different seasons and the character of the prevailing winds must all be considered as relevant. The main features of the climate, as determined by these factors and their interaction, will decide the important practical questions, not only of the kind of crops that can be grown, but also of the possibilities and scope for seasonal cropping.

Second only to climate and bound up with it, is the effect of soil on plant growth and crop husbandry. Soils are formed by the weathering of rocks, and although the mineral components are decided originally by the character of the parent rock, long periods of exposure to chemical and biological action show the influence of climate on the soil. In this way, under the same climatic conditions, soils may vary according to their origin, situation and age, while similar soil types may overlie rocks of different structure and composition owing to long exposure to a common climatic influence. Soil patterns can, therefore, vary considerably under similar climatic conditions, and the uniformity or variability in soil type can have far-reaching effects on crop husbandry.

The important characters of a soil are its physical structure, chemical composition, biological activity, and depth. The physical structure and chemical composition depend on the properties of the mineral matter and organic matter which compose the soil, and give to it the characteristic features as a medium for plant growth. These features are the 'texture', water-holding capacity, available nutrients, acidity or alkalinity, aeration and temperature. All of these natural soil characteristics can be modified within limits by management and cultivation, and although fertility is an inherent soil character, mechanical cultivation and the addition of inorganic and organic fertilisers have great effects on the capacity of a soil to grow crops. Systems of farming are characterised by particular forms of crop husbandry, and after the overriding influence of climate has played its part, the principal features of the crop husbandry are determined naturally by the soil type.

The physical features, climate and soil types of Britain make the country naturally suited to carrying broad-leaved trees on most of the land that is cultivable. With the destruction of the various kinds of woodland for the development of agricultural activities the farmer is provided with conditions that are suited to very characteristic forms of crop husbandry. The climate, although showing marked local variations is of a temperate maritime type free from extreme conditions, but with definite seasons through the year. There are usually no long periods when severe cold makes plant growth impossible, while the distribution of the rainfall throughout the year, and particularly during the long growing season, prevents extreme drought in the summer. The conditions are, in fact, suited to long periods of vegetative growth, and from the agricultural point of view the most important fact is that various types of grassland can be maintained artificially, and with little difficulty, over a considerable proportion of the land surface. This has resulted in both intensive and extensive use of grassland for agricultural purposes, and one of the most characteristic features of British farming is the development of a crop and stock husbandry based to varying degrees in different parts of the country on grassland.

But climate and soil also show their influence on British agriculture in the kinds of tillage crops and the methods of husbandry. A wide range of the more important cultivated crops of temperate regions can be grown successfully, and the conditions allow the land to be occupied by tillage crops throughout the year, according to the kinds of crops being grown and the rotation practised. There is also little difficulty in cultivating most soil types over a comparatively long period and at different times of the year, while the important operations of sowing and harvesting are not closely restricted with regard to time. This latitude is most important, and in addition to having alternative times for the sowing of some crops, as for example the spring and autumn sowing of wheat, barley and oats, the wide range of forage crops and vegetables that may be grown allow intensive cropping with good management and sound rotations.

For a country with such a small land surface, Great Britain shows surprisingly distinct regional differences in climate, and marked local variations in soil type, both of which are reflected in a general diversity of crop husbandry, a considerable range of crop rotation, and a characteristic distribution of tillage crops. In England, for example, there is a general association of wheat and barley with the lower rainfall areas, and of oats with the higher rainfall areas, while wheat is usually grown on medium to heavy loams, and barley on the lighter loams. The greatest acreage of sugar beet is in the east and south of England, and of swedes in the west and north, while grass leys are of longer duration in the cooler and wetter areas than under the drier conditions of the eastern counties. It is obvious that such distributions of different kinds of crops, when considered in relation to other factors like soil acidity, must lead to a wide range of cropping systems.

Although the natural growing conditions decide the main pattern and chief characteristics of the agriculture of a country, as well as determining the principal features of crop distribution and husbandry, it is seldom that there is a free interplay of natural conditions and cropping. This is particularly true when a country develops other important industries which employ a large proportion of the population with a consequent growth of a valuable internal and external trade in non-agricultural commodities. In such circumstances the agriculture that eventually stabilises has to adjust itself to a national economy that may be quite outside its control, and the influence of social and political considerations of a non-agricultural nature may be felt very strongly. The position may eventually be reached when the standard of living in the country bears little relation to the agricultural industry, and the activities and interests of the urban population become the dominating factor in the life of the country.

This is the position that has developed in Great Britain, and the whole agricultural industry is peculiarly placed with regard to the national economy and the feeding of the people. The country is over-industrialised and over-populated as a result of the prosperous days of expanding overseas trade in manufactured goods, and the output from British agriculture has become quite incapable of supplying the total needs of the people in certain staple foods like bread and meat. Combined with a demand for cheap food, there has been a great market in this country for a wide range of the more expensive foods and luxuries, and with plenty of money to be spent, a high standard of living, and little or no protection for the home-producer, food importation has developed on a vast scale.

The result has been that the British farmer has had to face very keen competition from overseas farmers, who in some countries have organised their production and marketing entirely for this country. The great agricultural expansion which occurred in many parts of the world in the nineteenth and early twentieth centuries, and the improved methods of transport and food storage which have made it possible to market agricultural products far from the centres of production, have maintained a constant competitive pressure on the British farmer. As a result of these developments, with changes in food tastes and requirements, the necessary seasonal food supplies for this country have been assured, as the producing capacity of various parts of the world are drawn upon. But the effect of this large-scale importation has been very serious for the farmer in this country. He has been unable to withstand the competition of cheap food produced under entirely different natural and economic conditions from those in this country, and in some cases the position has been made more difficult by other countries subsidising their exports of agricultural products. In an effort to meet this situation the British farmer has tried changing rotations and balance of cropping to find a system of agriculture that pays, while the whole balance of agricultural production in this country has been affected.

The problem that faces British agriculture is then not simply the organisation of a system of production that is suited to the natural conditions, but the method of adjusting itself to the complex economy of the country. This means that the crop plants cultivated in this country must conform to the requirements of an agriculture that is presented with the peculiarly characteristic economic situation of a nation primarily engaged in industry, commerce and overseas trade. The large and relatively well-paid urban population ensures a good market of high purchasing power, and this market obviously encourages the production on an intensive scale of the higher-priced and more perishable articles that are not easily stored or transported over long distances. The standard of living in the towns, and the constant labour competition, also tend to encourage larger farms, mechanisation and paid labour in contrast to the small peasant family farm of the poorer agricultural countries. All of these social and economic questions have played their part in moulding British agriculture, and the farmer's problem is to ensure a well-balanced system of husbandry that takes advantage of the natural resources and the home market.

Chapter III

BRITISH CROP HUSBANDRY

The dominant feature of the countryside in the lowland and upland districts of Britain is, except in certain special areas, the various types of grassland. It has already been explained why, because of the natural growing conditions, this should be so, and it is not surprising that the agriculture of Britain is very closely associated with grass. The total land surface of England and Wales is just about 37 million acres, of which approximately 30 millions are used in one way or another for agricultural purposes. In 1939, over 20 million acres were occupied by permanent grass and rough grazings, and only 9 millions were devoted to arable land including temporary leys. Permanent grass includes many different kinds of grassland of varying agricultural value and uses, but it has been the dominant feature of British agriculture for many years and occupied over 15 million acres in 1939. These figures, of course, fluctuate from year to year and during the two major wars of this century the area under tillage increased at the expense of the permanent grass. In the 1939–45 war, for example, the acreage of arable land reached over 14½ millions; the temporary grass increased by nearly 1 million; and there was a compensating decline in the permanent grass of 6 million acres.

The relative amounts of total grassland, and the proportions of the different kinds of grass, vary throughout the country according to the suitability of the conditions to different types of farming. The greatest proportions of grassland are found in the regions of high rainfall in the west, and as the climatic conditions become more suitable for arable crops, particularly cereals, in the eastern and southern districts, the amount of tillage land increases and grassland ceases to hold such a predominant position. Indeed, in the typically arable areas of eastern England, the arable land occupies a greater acreage than does the grassland, and the whole face of the landscape is changed by the work of the plough and the cultivation of annual crops.

But it is not only the major factor of climate that is responsible for this distribution of grassland and arable crops. Tillage land is obviously found in areas which are suited to work with agricultural implements both with regard to the nature of the countryside and the type of soil. Hilly and mountainous districts may be impossible or very expensive to cultivate, and although the availability of tractors and other mechanical implements has altered this situation to some degree, the soils in such districts are often shallow and infertile, and cultivation on steep slopes is liable to lead to soil erosion. Therefore, for economic and practical reasons, hilly and mountainous areas are usually left under grass, particularly as in this country, such areas are associated with high rainfall. Even under grass, it is difficult, and usually uneconomic, to look after this type of agricultural land as intensively as grassland under lowland conditions, and it is only to be expected that the highest proportion of rough grazings are situated in the high rainfall areas in the upland parts of the western half of the country.

Permanent grassland, which at its worst merges into rough grazings, and at its best is intensively managed and very valuable, is distributed throughout the country under many different conditions. The very extent of this type of agricultural land means inevitably that it dominates farming practice over a large part of the country, and its status with regard to productivity has far-reaching effects. The best types of permanent grass are found in lowland areas of good soil fertility: they are associated with good farming conditions and are the centres of intensive livestock husbandry, but are unfortunately a small proportion of the total permanent grass. By far the greatest amount of permanent grassland is indifferent or poor in quality, although varying considerably in different localities and under different systems of management. It occurs throughout England and Wales, and becomes poorer at higher elevations and on infertile and shallow soils, eventually merging into the rough grazings with neglect and impoverishment. It must be remembered that grassland in this country, though first class and very valuable when managed properly, will not maintain itself when neglected, and will revert to scrub, woodland and other useless vegetation from the agricultural point of view.

The extent and the distribution of grassland of all types in this country naturally means that, on this judgement alone, the agriculture is centred on the keeping of livestock. Grass, whether grazed, cut for hay, dried or used for ensilage is essentially a livestock food, and British agriculture in normal times is mainly devoted to the production of livestock products. When considered as a crop, therefore, grass in all its great range in quality, methods of management and form of utilisation is easily the most important crop product of the country. But grass must be considered from another aspect in addition to its direct use as a food for livestock. Land under grass that is well managed is not only being maintained in fertility, but is being subjected to improvement. Good grassland husbandry not only ensures a high stock-carrying capacity, but it is a means by which soil fertility can be raised so that the land can be made to yield heavier crops when the grass is broken and arable crops taken. It is for this reason that farmers are being urged to regard grass as a crop which needs good husbandry, and that temporary leys should replace permanent grassland wherever this is possible, and become the pivotal crop on the farm.

But the concentration of crop husbandry on producing food for livestock does not stop with grassland. The greater proportion of the crops cultivated on arable land are grown for stock feed and no other purpose, while some crops are utilised partially for human consumption and partly for stock. Even certain crops whose utilisation is primarily for human consumption have valuable by-products that help to maintain farm animals. There is, in fact, only a small proportion of the total crop production of this country which is used directly for human food, in spite of the fact that certain vegetable crops have in recent years tended to pass into the hands of the farmer instead of being confined to market gardens. The idea of British agriculture being mainly concerned with growing crops for direct human consumption is, therefore, not correct, and it is only by the full appreciation of this fact that the whole structure of the agriculture of the country can be appreciated.

In spite of the dominating position of permanent grassland, the true character of rotational farming is decided by the utilisation of the arable land, which is the principal feature of the more intensively farmed parts of the country. The proportion of the total agricultural land which is treated as arable, and therefore comes under the plough, is not constant from year to year, nor is it the same in different parts of the country. Taking the country as a whole, the amount of arable land is a fair indication of the intensiveness of food production, and of the economic position, not only of agriculture, but of the country. Increased pressure of food production, and a greater demand for plant products for direct human consumption, always leads to an increased·acreage under the plough because more food per acre can be produced in this way.

Arable land, which in 1939 was roughly one-third of the total acreage under crops and grass in England and Wales, includes temporary grass and tillage. Land in tillage is carrying crops other than temporary grass, or is being bare-fallowed in preparation for such crops, and the greater proportion of the arable land in England and Wales is usually being used in this way. Temporary grass is included with the arable land because it is part of the rotation on the farm: it is essentially cultivated land, the grass being regarded by the farmer as a crop occupying the land for a varying but limited period according to the particular rotation being used.

The complete picture of the way in which the agricultural land is being used can only be seen when one can visualise how the tillage acreage is divided amongst the important crops. It is here that the characteristic feature of rotational farming in this country is evident, and the fields of corn that are traditionally associated with the plough assume their correct proportion, which was actually nearly 60 % of the tillage in 1939. The three cereal crops, wheat, barley and oats, grown as pure crops, and not as mixtures, make up this total, wheat occupying 25 %, oats 20 % and barley 13 % of the total tillage land, although a small proportion of mixed corn and rye is also cultivated.

Second only to the corn crops are the various 'roots' which are also so characteristic of the husbandry of this country. These

2-2

crops, in which may be included potatoes, mangolds, swedes, turnips, sugar beet and fodder crops such as kales and cabbage, occupied a further quarter of the tillage. These figures may be analysed one step further: the two crops used directly for human consumption—potatoes and sugar beet—were together approximately equal in acreage to the roots and fodder crops grown for stock feed, and when to this is added the acreage under fruit and vegetables (8 % of the tillage) there is a total of only just under 20 % of the total tillage devoted to crops which are used directly as human food.

There still remains several miscellaneous crops such as beans and peas for stock feed ($2\frac{1}{2}$ %), vetches, flax, linseed and hops, which together with the area being bare-fallowed to clean the land, complete the total tillage acreage. Although some of these miscellaneous crops, as for example hops, are very important and of relatively high cash value in certain restricted areas of the country, they are not dominant features of the crop husbandry on the tillage land. The cereals and the roots are the mainstays of rotational practice in this country, and it is around these crops that we find the farmer planning his systems of intensive cultivation wherever the conditions are suitable.

The position in Scotland is different from that in England because of the nature of the country and the climate. As in Wales, there is a high proportion of land which is quite unsuited to intensive agriculture, and only about one-quarter of the total land surface is under crops and grass, excluding rough grazings. But there is normally more land devoted to arable farming than to grassland in Scotland, and in 1939 the proportion was two-thirds arable to one-third permanent grass, the reverse position to that in England and Wales. It is also characteristic of Scottish farming, that temporary grassland occupies a relatively high proportion of the arable acreage, when judged on standards in England.

The use of the tillage land in Scotland shows the important effect of climate. Although nearly two-thirds is used for growing corn, wheat and barley together contribute only a small amount, and oats are easily the most important crop, because of all the cereals oats are most suited to cool and moist conditions. Now oats are essentially a stock feed, and the large amount of this crop, with the intensive cultivation of temporary grass, enable the Scottish farmer in the lowlands to concentrate his efforts on keeping stock, and it is natural that the other important tillage crops should be swedes and turnips. The only really important tillage crop grown in Scotland for human food, apart from the proportion of oats used for oatmeal, is the potato, and although potatoes are grown on a higher proportion of the tillage than in England and Wales, they are much less extensively grown than even swedes and turnips.

Although the proportions of the agricultural land devoted to different crops varies in different parts of Great Britain the general picture is of a country using the greater part of its available resources for growing food for livestock. The greatest acreage, taking the country as a whole, is occupied by permanent grassland, although in certain areas there is very little land used in this way. A proportion of the arable land also carries grass, but of a temporary nature, while the tillage crops are varied and are grown both for human and livestock consumption. Of the tillage crops, the three cereals—wheat, oats and barley—occupy the greatest acreage and, with the various roots such as potatoes, swedes, turnips, mangolds and sugar beet, are the characteristic features of British agriculture.

Chapter IV

THE ROTATION OF CROPS

It has already been stated that it is characteristic of the crop husbandry of Britain for the land to be used to grow a considerable range of crop plants. Each farm develops its own cropping system, which of course may be common to many farms, and is so planned that the same tillage crop is usually not taken from one and the same field year after year. In some cases it is the custom to adhere to a more or less rigid system of cropping, while in others the system is more flexible and the farmer allows himself certain modifications from year to year as the occasion may require. Whatever system is adopted, the general plan of cropping, if it is to be efficient, must be devised to fit in with the growing conditions, the markets, the general economy of the farm, and the tenets of good husbandry.

Any sequence of cropping, or fixed rotation, is then the result of a careful adjustment of methods to the circumstances, and is not a haphazard means by which the farmer can grow several different crops on the one farm. The development of cropping systems in this country became necessary in the early days of settled agriculture and arable farming, to allow the land to be kept in tillage without resulting in the soil becoming unfit for cultivation and bearing crops. Cropping systems were at first simple, when land was plentiful and the available crops few in number, but with the increased pressure of population and more intensive food production, rotational systems became more efficient and more complicated. But although more crops were taken into cultivation, it also became customary in some cases to follow peculiar systems with comparatively few crops. Revolutionary changes in rotational practice followed the introduction of roots and red clover to this country in the eighteenth century, and standard rotations based on a fixed sequence of roots, barley, grass and clover, and wheat, were developed in most parts of the country. There was, of course, considerable variation in the practices followed in various parts of the country, but the rotation of cereals, root crops and clover was widely practised, while in some areas rotations were devised specifically to introduce peas to improve soil fertility. The advantages of such a system of cropping became so widely recognised, that tenant farmers were forced by law to farm according to the requirements of the landlord, and it was not until the beginning of the twentieth century that freedom of cropping for tenant farmers was finally achieved.

The modern tendency in rotational practice is for a greater flexibility in the choice of crop sequences and a less rigid adherence to tradition. There are many reasons for this change, the more important being the unstable economic conditions, the wider and more general application of mechanised equipment, the increased use of artificial fertilisers, and the greater emphasis that is being placed on the inclusion of the temporary ley in the rotation. In some cases the introduction of new crop sequences is nothing more than a desperate attempt on the part of the farmer to adjust his methods to changing economic conditions, and the whole balance of cropping and farm output may be altered. Often such attempts are not based on good husbandry, and can only be regarded as temporary emergency measures to tide over bad times. However, many innovations in cropping, although perhaps forced by economic circumstances, are the results of sound enterprise and shrewd originality on the part of the farmer, and they have done a great deal to increase the efficiency of crop production and to adjust farming systems to changing times and opportunities.

Economic considerations and specific market demands are very important in their effects on rotation, but even if there was no such complication it would still be necessary for the farmer to adopt some planned cropping system for the permanent and efficient cultivation of his farm. In other words, cropping sequences have not been developed solely as the result of changing economy or public tastes and requirements, although these are extremely important. The first essential of a good rotation of crops is that it should be based on sound principles of husbandry, and these principles cannot be flouted in any permanent and intensive agriculture which depends on a high level of production for its survival as does the agriculture of this country.

The foremost consideration for the farmer is obviously to maintain, and if possible to improve, the productive capacity of his land, and the basis of all crop rotations is the proper care of the soil. Some soils are inherently of a high fertility, while others are not, but whatever the inherent characters it is necessary to maintain the soil in a condition which is economically productive under the particular circumstances. A fertile soil in a good condition has a readily available supply of mineral substances for plant nutrition, a good physical texture suitable for cultivation and as a medium for root growth, an active micro-organism population which brings about desirable chemical changes, a sufficient organic matter content, a suitable acid-alkali balance which prevents undue acidity or alkalinity, and an absence of weeds, insect pests and fungal diseases. Although the physical and chemical nature of the soil are fundamentally natural attributes, they can be modified within limits by the agriculturalist, while cleanliness of the land in terms of relative freedom from weeds, pests and diseases is largely a matter of good husbandry.

Sound rotations, from the husbandry point of view, are then directed towards keeping the arable land productive and maintaining or improving its fertility by due attention to all the contributory causes mentioned above. Rotations make this possible because of the individual requirements, methods of cultivation, effects on the soil, relationship with pests and diseases and form of utilisation of individual crops. Rotational practice ensures that not only can the various crops be grown on different fields each year, but also that each individual field is subjected to the benefits of cultivations and manuring which are associated with the preparation of the land for sowing the crop and controlling weed growth. Rotations do not, of course, imply that every field bears a different crop each year, but they do ensure that the arable land carries a balanced sequence of cropping that prevents soil deterioration in one form or another, and this is in the main achieved by alternating soil exhausting crops with those which allow for recuperation.

The operation of rotations in maintaining soil fertility and condition may be exemplified in several ways by considering in more detail the contrasting implications of continuous cropping and a balanced sequence. Individual crops have different manurial requirements, and they remove from the soil characteristic amounts, both absolutely and relatively, of the nutrients present in the soil solution. Consequently, if the same crop is grown on the same field for a number of years in succession, the soil may become so depleted of certain nutrients that crop yields will suffer. On the other hand, if different crops are grown in a well-planned sequence this danger is considerably reduced, particularly when it is remembered that each crop has a characteristic root system and absorbs most of its mineral requirements from a particular level in the soil, according to the depth of root penetration and the distribution of the actively absorbing root branches. Crop root systems are extremely important in their effects on soil fertility by virtue of their very complex action which is mechanical, chemical and biological and not merely a simple withdrawal of certain minerals.

The mineral nutrition of individual crops varies, and it is necessary to make good the depletion of the soil by adding organic and inorganic manures if the soil fertility is to be maintained. But it is not possible to manure all crops in the same way and to the same degree, and the advantage of rotations is that the necessary soil applications can be made according to the crop which is occupying the land. For example, while it is possible to build up soil fertility by adding heavy dressings of dung or fertilisers to root crops, it is not practicable to do this with cereals which would probably become 'laid'. Moreover, some crops are more exhaustive for the soil than are others by reason of their utilisation. When the product is sold off the farm there is a complete loss of the soil nutrients which fed the crop, but when some or all of the product is fed to livestock on the farm, there is a residue which can be returned to the land as manure.

Although the main concern of the farmer is to maintain or increase the nutrients in the soil, while at the same time preserving the physical texture, this is not sufficient to keep a cultivated soil in a high state of productivity. When land is cleared of its vegetation, and cultivated for crops, there is always a strong tendency for it to be colonised by miscellaneous plants, which because they are unwanted, are classed indiscriminately as weeds. It is possible, by continual and intensive cultivations, accompanied by sound management, to prevent weed colonisation reaching such a serious condition that it interferes with crop growth, or hinders cultivation and harvesting of that crop. Weeds are an inevitable result of tilling the soil and cultivating crops, and it is impossible to eradicate permanently the unsown plants of cultivated land, because the predisposing conditions to invasion and colonisation follow naturally from the circumstances. The weed problem must be accepted, therefore, as a challenge to good husbandry, and the agriculturalists' task is to find the most economic solution.

The various weed-combating, mechanical operations that have been devised are still the most important and widely used means of keeping weeds under control, and the rotation of crops gives the widest scope for using these operations. If the same crop is taken off the same field year after year, the land becomes foul with characteristic weeds which are suited to the soil, the management and the habit of the crop. The only way to reduce the weeds, except by chemical means, is to cultivate other crops which allow the land to be cleaned at different times of the year, or over longer periods, than was possible when the original crop was being grown. Winter wheat, for example, occupies the land for a long time and does not offer much opportunity during the growth of the crop for cultivations to destroy weeds. Weeds germinating in the autumn in winter wheat have the opportunity of becoming well established, and unless drastic spring cultivations are possible, the weeds grow away with the crop and set seed before harvest. A spring-sown root crop, on the other hand, not only gives an opportunity for thorough cultivation in the spring before the crop is sown, but it also allows cleaning operations during the early part of the development of the crop. It should, however, be emphasised that cleaning crops only perform their right function if properly managed and well grown. Owing to the different habits and life cycles of the various species of weeds that infest cultivated soil, it is necessary to adjust the cropping so that the fields on the farm can be subjected to weed-controlling operations at different times of the year. The inclusion of 'cleaning crops', which allow the maximum amount of cultivations to be done, is usually the only practicable means of keeping the weeds under control, and the alternation of cleaning crops with others sown in the autumn and spring, is the usual practice.

The other aspect of land cleanliness is freedom from fungal diseases and insect and other pests. The cultivation of a susceptible crop, either continuously year after year, or too often in a rotation, gives the parasite an opportunity to multiply to such proportions that it may cause serious losses in the growing crop. Such local building up of the parasitic population can also become a centre of infection, not only to other fields and crops on the same farm, but also to neighbouring farms and more distant areas if the diseased crop is transported any distance for consumption. One of the commonest means of progressive parasitic infection is by the soil becoming increasingly populated by an organism which relies on the crop plant to build up its numbers, but which can persist for years without the host plant. The practical remedy for this is to avoid growing the susceptible crop too often, and to plan rotations so that crops that are not affected occupy the land successively year after year for as long a period as possible. This method of control is used for eelworms of sugar beet and potatoes, which have become so serious in some areas that special action has to be taken to avoid growing the same susceptible crops too frequently. Clover stem rot and 'take-all' disease of wheat and barley, both of which are soil-borne fungal infections, can similarly be kept in check by suitable rotations, while rotational practices which control those weeds which may harbour certain parasites of crop plants play an important part in reducing fungal and insect depredations.

A planned sequence of cropping should then concentrate on keeping the farm in as high a state of productivity as is possible, simply by maintaining soil fertility and preventing the multiplication of pests and diseases. But economic considerations with regard to labour and the production of marketable produce also play an important part in determining rotations and the type of farming. Crops are seasonal in their labour requirements, and it is impossible to utilise efficiently the man, horse and implement power if the work is crowded into a few peak periods of the year, with long intervals of inactivity in between. On this basis alone, and particularly when considered also in relation to the seasonal use of the land, rotations are a virtual necessity.

Farming systems which include the production of more than one marketable article obviously involve less risk than those which specialise on one product. Market demands and prices vary from season to season, while weather conditions are notoriously variable during the growing season in this country. With rotational cropping the economic failure of one crop may be compensated by the success of another, and where there is a special local demand for particular products, or a climatic and soil adaptation for the growing of a certain crop, the principal characteristics of a rotation may be decided by these considerations.

In the final analysis, the crop rotation must contribute its share to the profits of the farm, and a good rotation must not only be good husbandry, but also sound business. The rotation must be adjusted to the whole undertaking of the farm, and thereby combine efficient farming with profitable production. It may either supply the bulk of the cash income, or be subservient to the demands of a livestock husbandry on which the farm depends for its marketable produce.

Rotations are necessary, for example, to provide a seasonal supply of forage and the different kinds of home-produced food for the various classes of livestock on the farm, and carefully planned cropping systems may be devised entirely for this purpose. Therefore, the cropping balance and sequence will depend on the particular agricultural undertaking with special reference to the emphasis that is being placed on the direct production of human food or the conversion of plant products to animal products.

Chapter V

GRASSES AND GRASSLAND

Various types of grasslands are not only the most extensive and characteristic kind of agricultural land in Britain, but they occupy a larger area than any other form of vegetation. These grasslands owe their origin and their maintenance to different sets of circumstances, they differ in their appearance and in the kinds of plants which compose them, and their value and utilisation by the farmer vary enormously. Grassland is, of course, not a product of grasses only, but other valuable plants contribute to its botanical and agricultural characteristics, while it is seldom that worthless or even harmful plants are not found entering into its composition. This, however, does not materially affect the position that grass is the most important single crop grown in this country, and this by virtue of the fact that the growing conditions favour grassland, not only as a crop, but as a vegetation type also, over a very high proportion of the agricultural land. It is only natural that British agriculture should have exploited this, and by taking advantage of the natural circumstances, developed a husbandry which, in many of its features, is essentially of a grassland type.

The great economic value of grasses makes it worth considering their characters as a botanical group of plants, and examining them in relation to their extremely important agricultural uses. With the cereals, the grasses belong to a very large and characteristic family named the Gramineae, which is distributed throughout the world and includes about 500 genera and thousands of species. This family is by far the most important of any family of plants as far as agriculture is concerned, and is responsible for supplying staple foods for man and animals in all parts of the world, as well as being the source of some important plants which have long been used for building and constructional work in tropical and subtropical countries. It is no exaggeration to say that agriculture over a large area of the earth's surface has grown up on graminaceous plants, and most systems of husbandry are primarily dependent on grasses or cereals for the direct or indirect production of staple foods.

The grasses are the outstanding animal forage crops, and their virtues for this purpose are due to their peculiar morphological characters, and the great range of forms adapted to a wide range of conditions which exist in various parts of the world. The first and most obvious character of grasses is their habit and mode of growth by which they form close tufts of edible leafy shoots close to the ground. These leafy shoots, or 'tillers', are the only form of branches, with the exception of the flowering stems, that most grasses normally produce. Some species, however, develop creeping stems below or at the surface of the soil by means of which they can propagate themselves and colonise new ground. This habit, and the normal method of vegetative branching, give grasses the extremely valuable property of being able to form a continuous cover to the ground, a property which, however, is not shared equally by all species.

Owing to the fact that all the branches are produced at ground level, and the growing tips of these branches are protected at the bases of the leaves while in the vegetative state, perennial grasses can be cut or grazed periodically without doing permanent harm to the plant. In fact, removal of some of the leafy shoots stimulates the production of more branches, and when the leaves are cut or grazed while they are actively growing, they can continue to make more growth from below. Even when the flowering stems are cut off near the base, the perennial forms (and some annual forms if cut sufficiently early) are stimulated to make fresh growth by producing more tiller branches. In this way most grasses are protected from excessive damage through the removal of the flowering stems by the presence of basal leafy shoots which do not grow up and produce inflorescences, but which continue the vegetative growth of the plant. The proportion of vegetative shoots to inflorescence shoots varies in different species and strains of grasses, and is a valuable agricultural character.

These qualities of grasses make them peculiarly suited to carrying grazing animals, and also to periodic cutting, although some species and strains are more suited to the one form of treatment than the other. The important practical point is that the types which are suited to grazing not only withstand the action of grazing animals, but they flourish and improve in growth under skilful farming management. Grasses are, in fact, the ideal grazing plants: they can withstand treading and defoliation by stock, they crowd out weeds and they form a continuous ground cover under good management. Their response to skilled management is seen in improved swards which give higher yields of more nutritious animal food over a more sustained period of growth.

It is, then, as herbage grazing plants that grasses are pre-eminent, and it is in this way that the greater part of the grass-land of this country is utilised. But grass is equally valuable as a fodder plant which can be conserved as hay, silage or dried young grass to supply livestock food at seasons of the year when grazing is impossible, or fresh grass is in short supply. The same grass field is often used for both grazing and cutting for hay, but it is usual to distinguish between the two forms of utilisation when considering the agricultural types of grassland in the country. In England and Wales approximately three-quarters of the permanent grasslands are managed as grazings, while less than half of the rotational grassland is treated in this way. Such an apportioning of the grazing and hay lands between the two main types of grassland recognised agriculturally is only to be expected from the distribution of the permanent and temporary swards and their association with particular kinds of farming.

The grasses occupy a peculiar position as cultivated plants among the valuable agricultural plants which dominate farming in this country. Of the many grass species which are found in Britain, only a very small proportion has any agricultural value, and the species that occupy the greatest area owe their value simply to the fact that they occur so widely, and not because they have the greatest feeding value or can support the largest number of stock to the acre. Such grasses are not cultivated in the strict sense of the term; they have for the most part colonised naturally land which was originally cleared from other vegetation, and they maintain themselves with the minimum amount of attention from the farmer. These grasses have not been subjected to improvement by the grower or by plant breeders, but they are accepted as useful for want of anything better which will fulfil the same function under the particular circumstances. The most widely occurring grasses of this type are the various bent grasses, which are species of *Agrostis* and colonise practically all forms of permanent grass in this country. They are, in fact, the typical grasses of the 'uncultivated' grasslands, although they owe their survival to farming management and grow extensively on enclosed grazings ranging from upland and mountain areas and sheep walks to intensively managed permanent swards in the lowlands. They can exist under poor conditions and management, and are typical of the 'low fertility' agricultural grasses.

At the other extreme are grasses such as the rye grasses, timothy, cocksfoot and rough-stalked meadow grass. These are grasses of much greater potential value than the bent grasses, and are typical of many of the best grassland areas in the country and also of the rotational temporary swards. They thrive under good management, on fertile soils and with high farming. Some of them have been subjected to intensive efforts in plant breeding, and all have been consciously encouraged. Seeds of these grasses have been collected, and areas set aside for seed production, and they have achieved a position as cultivated plants comparable to many tillage crops of much greater age under domestication.

Apart from the exceptional case of temporary rotational leys which have been sown for special purposes, grass swards are always composed of more than one species of grass, and the value of grassland in terms of productive capacity is dependent on the kinds of species that dominate the sward. Grass which is of a permanent nature, or which is down for long periods of time, gradually takes on a characteristic botanical composition which is a reflection of the growing conditions and the way in which it is managed. There is no other agricultural crop which is exposed to such varied conditions of soil, climate and management as is grass, and certainly no crop is so intimately and persistently at the mercy of the environment in which it grows. Consequently, it is possible to distinguish different kinds of permanent grassland

PLATES

1–9

GRASSLAND ASSOCIATED WITH WOODLAND—A FEATURE OF BRITISH AGRICULTURE

PLATE I

The association of grassland and woodland is one of the commonest features of the landscape of the British Isles, and grassland occupies a greater area than any other type of vegetation. Cleared woodland can grow excellent grass when properly tended and managed, and the growing conditions of this country are particularly suited to grassland husbandry. Although grassland is by far the most extensive type of agricultural land in Britain, there is a great deal that is poor and indifferent in addition to that which is intensively managed and of good quality. It is, however, the suitability of the growing conditions for maintaining grassland and raising livestock that has resulted in British agriculture developing its characteristic grassland husbandry as the basis of the livestock industry which is such an obvious feature of the agriculture. The young stock in this illustration are grazing a reseeded pasture on typical parkland, and it may be noted that these mixed cattle do not comprise a single herd of one breed.

PLATE 2

The grasses are the most important cultivated plants of British agriculture, and the economic value of grassland depends on the botanical composition of the herbage. All grasses belong to the botanical family Gramineae, and there are many genera, species and forms composing this family, although only a comparatively small number has any great agricultural value. Where the conditions are suitable, the farmer can establish and maintain first-class grassland by encouraging the growth of the best grasses either by reseeding or by proper management. It is possible, even in the same field and under similar growing conditions, to have entirely different grass swards growing side by side. Although skilled and intensive management can do a great deal to improve swards which have deteriorated and become rough, the most satisfactory method of improvement is to plough out the old sward and reseed with a suitable mixture of good grasses.

SOWING A GRASS LEY ON SPECIALLY PREPARED LAND

PLATE 3

The development of 'artificial' grassland, and particularly the temporary ley, is the most valuable method of using grasses under intensive systems of agriculture. The best swards can only be established if the land is first properly prepared for sowing, and by the proper choice of the most suitable species and strains of grasses, the farmer can establish the kind of sward he desires. The sowing of grass seeds mixtures for cultivated leys, and the incorporation of these leys in the farm rotation, not only leads to improved grassland and the most efficient use of grass as a crop, but is also the means by which soil fertility can be maintained and improved.

A PURE STAND OF CLOVER TYPICAL OF THE HAY TYPES

PLATE 4

The clovers are the most valuable leguminous plants cultivated in Britain. They are used in grass seeds mixtures for all types of grassland, and are valuable for both grazing and hay according to the species and strain used. The larger growing and bulkier species such as red clover, crimson clover and alsike are used for hay, while the white clover and wild white clover are used for grazing. The red clovers are also used in very simple mixtures, or sown by themselves, for cutting green or making into hay, their agricultural value depending on their capacity to produce large amounts of nutritious herbage. Between them, the available species and strains of clover offer a wide range of types suited to various forms of management and kinds of growing conditions.

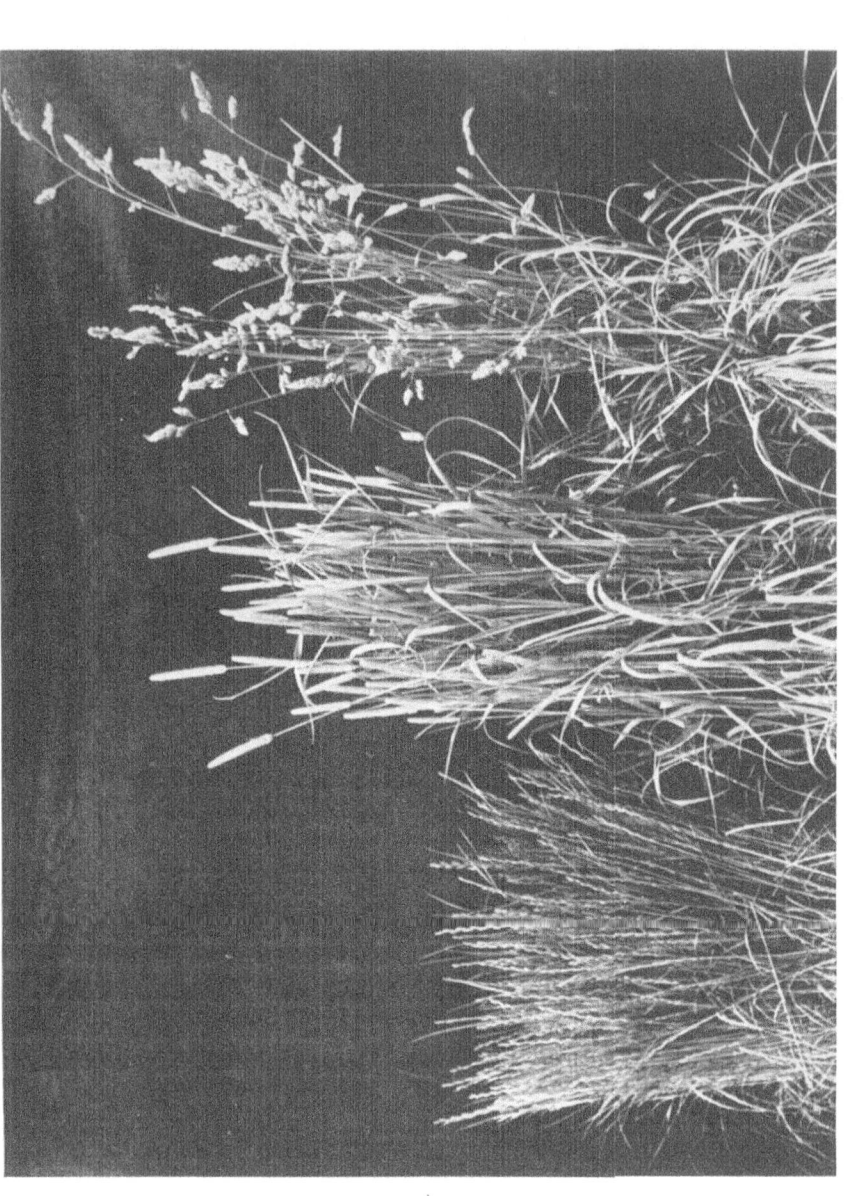

THREE IMPORTANT CULTIVATED GRASSES FOR ESTABLISHING 'ARTIFICIAL' GRASSLAND

PLATE 5

The three most important 'cultivated' grasses in British agriculture for the sowing of temporary and permanent swards are cocksfoot, timothy and perennial rye grass, each of which belongs to a separate genus of the family Gramineae. These grasses are widely distributed in the temperate regions of Europe and Asia, and a large number of different forms of each exists. The most significant development in grassland husbandry during recent years has been the recognition of these different forms or strains, and the selection of strains suitable for particular growing conditions and forms of management. A high proportion of seeds mixtures for agricultural purposes contains one or more of these three grasses represented by one or more strains suitable for the type of sward which the farmer requires.

TWO STRAINS OF WHITE CLOVER

PLATE 6

White clover has been described as the most important cultivated plant of British agriculture because of the vital part it plays in the composition of many different types of grassland and its use in grassland improvement. Of all leguminous herbage plants, white clover is the best adapted to grazing purposes, its low growing and creeping habit enabling it to withstand treading and defoliation by livestock. There are two main types available to the farmer—the ordinary, Dutch or cultivated white which is larger growing, earlier flowering and shorter lived: and the wild white which is finer leaved and more or less permanent under suitable management. Many strains of each type are on the market, and the true characters of these strains for the certification of stocks, is determined by growing small observation plots side by side and comparing the behaviour.

A LUCERNE LEY

PLATE 7

Lucerne, although not cultivated extensively in Britain, is the world's most important leguminous forage crop for temperate conditions where it is most valuable under warm, low rainfall climates. In this country it is grown mostly in the south-east, where its drought-resistant characters are particularly useful. Lucerne may be sown alone or in simple mixtures with one or two grasses, and it may be grazed, made into hay or silage, used for green soiling or making into lucerne meal. If the conditions are suitable, the crop will remain productive for several years, and it may be cut three or four times in one season. There are many strains and types of lucerne, and the choice of the most suitable strain is the most important consideration in successful lucerne cultivation.

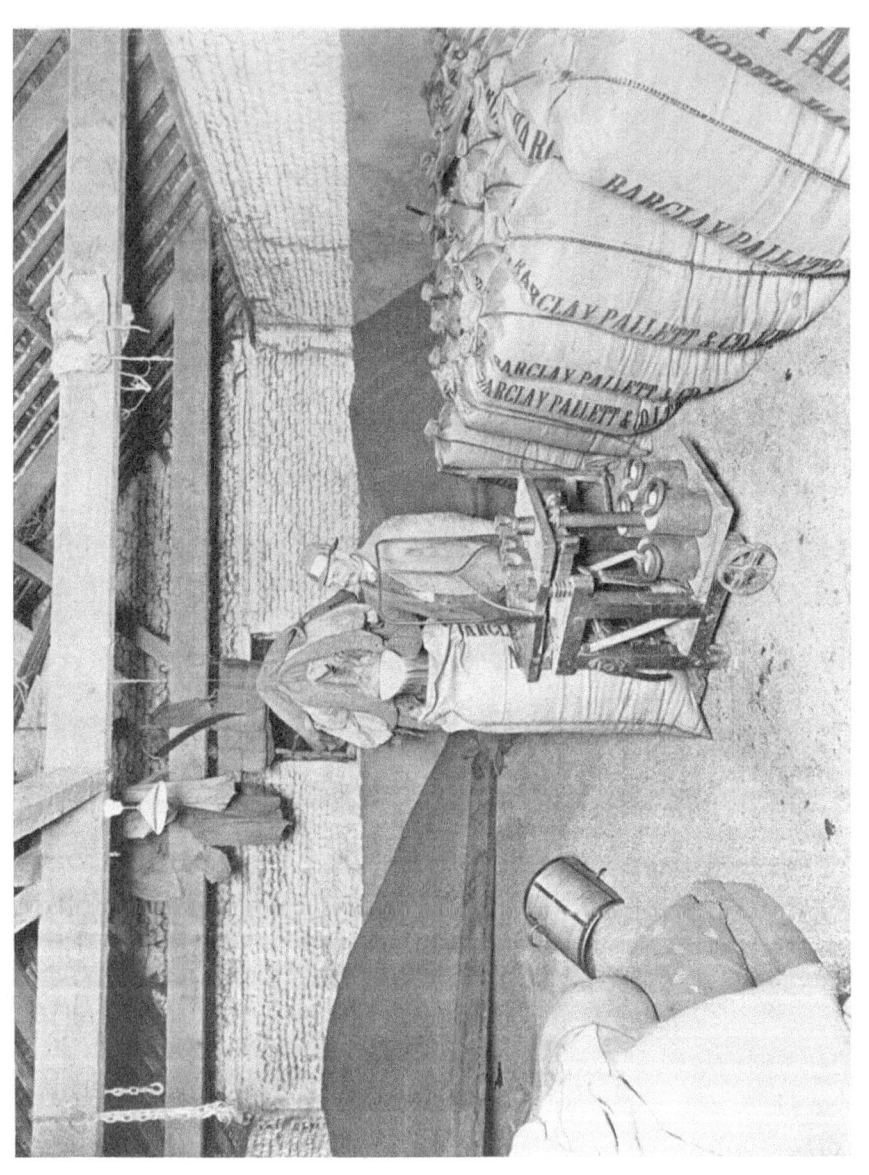

THE FARM GRANARY

PLATE 8

The cereals, which are the grain producers of the family Gramineae, are the world's great providers of human food, although vast amounts are also used for stock feed, and a smaller proportion for making fermented drinks. The cereals have been the basis of crop husbandry and food production since the beginning of settled agriculture of the oldest civilisations, and the development of cereal cultivation is the dominant feature of world agriculture to-day. Although cereals are characteristic crops of most farming rotations in this country, and high yields of wheat, barley and oats can be grown under suitable conditions, Great Britain is not an important grain-producing country judged on world standards.

WHEAT IN STOOK

PLATE 9

Wheat and rice are not only the most important cereal crops of the world, but these two cultivated plants provide the staple diet of more people than any other food. Wheat is the world's pre-eminent bread corn, and is grown more extensively than any other crop at the present time. This country cannot grow sufficient wheat to meet the demands at home even for bread, and this alone makes it necessary to import the greater proportion of the wheat consumed, in spite of the high yields that are obtained in the good wheat-growing areas of the country situated in the low rainfall and sunnier districts. Although the baking quality of wheat varies considerably between varieties, home-grown wheat has certain defects for bulk handling and breadmaking. The grain is therefore used principally for blending with imported wheat for making various kinds of flours, as well as being used for confectionery flour, biscuits and poultry feed.

in this country according to soil, rainfall, topography and local peculiarities of conditions. In any area the effect of the growing conditions can be very much modified by the agricultural management of the grassland, so that two fields adjacent to one another may have quite different grasses dominating their swards.

To understand the grasslands in this country and their relationship to agriculture, one must appreciate the various kinds and types which exist under different conditions. Some authorities on grassland recognise two main types—the un-cultivated, which includes a large part of the rough and hill grazings, and the cultivated, which is the more important because it includes all the lowland permanent pasture and meadowland as well as the temporary leys. It can be seen that a division of this kind is quite artificial and not very easy to put into practice, because in the first place it is not always a simple matter to decide what is cultivated and what is uncultivated, and in the second place long leys merge into permanent grass. But if we accept a grouping of the grasslands in this way there is no doubt that it gives a useful starting point from which to consider the grassland types of the country.

The rough and hill grazings, which are the most important part of the uncultivated grassland of Britain, are found most extensively in the west and north of the country, and much of it is typical moorland on peaty, acid soils. The vegetation is very mixed, there being rushes, bilberry, bracken and other plants growing in varying amounts with the grasses according to the depth of the soil, the drainage and other natural conditions. The grass species themselves also vary with the conditions, the commonest and most important being the bents (*Agrostis* species), fine fescues, mountain moor grass and mat grass. None of these is a really valuable agricultural grass, and with the roughness of the grazings and the varying amounts of other plants which are hardly eaten by stock, these areas are of a low productivity.

In the south and east of England there are other types of rough grazings occurring on the Down and dry heathy areas. These are of a quite different type from those of the north and the west; they sometimes have scrubby thickets and are often infested with rabbits, but they resemble the other types of rough grazing in that they have a very poor stock-carrying capacity. All forms of rough grazing indicate poor farming conditions, and sometimes even downright neglect or impoverishment due to uneconomic conditions for their proper management, and they represent as a whole a type of agricultural land which must be kept in check.

The cultivated permanent pastures represent the largest group of all the grassland types in Britain. Their botanical composition varies considerably, and their agricultural value depends on the proportion of valuable herbage plants such as perennial rye grass and wild white clover. Unfortunately, however, a large proportion of these pastures is of inferior value because the botanical composition consists largely of poor pasture plants, particularly the various bent grasses. The best of these permanent pastures, composed of the highest amounts of perennial rye grass and other valuable species, occur on fertile, lowland soils which are associated with intensive livestock industries like dairying or bullock fattening. The standard and economic value of these pastures depends to a large extent on their management, the best of them owing their excellence to skilful treatment by the farmer. It is obvious, therefore, that the really first-class permanent pastures occur in isolated patches in various parts of the country, and they are often found amongst very inferior pasture which is suffering from neglect or mismanagement, and has large amounts of bent grasses. These bent grass, or *Agrostis*, pastures are the most extensive of any of the grazing swards in Britain, and are found as the typical type on acid soils. Many of the old and neglected permanent pastures on heavy soils are composed largely of bent grasses, even in dairying districts, and some of them have been used for carrying cake-fed stock and have served the purpose of not much more than convenient exercising grounds for the beasts. Where the management is very bad, and on the heaviest and wettest soils, this type of sward becomes rough and colonised by even more inferior grasses and by rushes which may gradually encroach on the field and make it practically worthless. At higher elevations the *Agrostis* pastures often become colonised with fine fescues, particularly where the soil becomes more acid, and the sward suffers in consequence.

It is interesting to see how different types of permanent pasture are found associated with soils of varying acidity and alkalinity so that it is possible to classify grasslands as acidic, neutral or basic. On acid soils the most extensive type is the bent grass and fine fescue sward which is found on the lower slopes of mountains, on hill sides, and on the heaths and commons in the lowlands. Where the soil becomes peaty, the mat-grass type comes in, and in wetter areas where the soil is boggy, the moor grass or flying bent becomes the most prominent grass species. Neutral grassland is found where the soil is neither very acid or very alkaline, such as lowland loams and clays, and is characteristic of the many types of lowland permanent swards. The number of grasses which grow under these conditions is greater than on the acid soils, and includes such valuable grasses as perennial rye grass, cocksfoot, timothy, meadow foxtail, rough-stalked meadow grass, meadow fescue and others of lesser value like crested dogstail and sweet vernal grass, with the ever-present bent grasses in varying amounts. On the chalks, limestones and other basic rocks, are developed some of the healthiest pastures dominated by sheep's fescue and red fescue, with other grasses such as the erect brome and the false field brome.

It needs no further emphasis to show what a variable crop is permanent grass, nor to point out how misleading it is to think of this type of grassland, or any other, as a uniform agricultural crop product like wheat or potatoes. Grassland is a great national and agricultural asset which responds to good management and treatment to an amazing degree. It can be virtually worthless under poor conditions and neglectful husbandry, but there is much of it that can be improved out of all recognition if it is economically worth while to do so. Improvement is possible only by changing the growing conditions through soil amelioration and good management, which together encourage the growth of the better grasses. First-class grassland cannot be obtained with poor species of grass, and the best species of grass can only be grown successfully under good conditions.

3-2

There has been considerable discussion in recent years as to the relative merits of permanent and temporary or rotational grassland, and the extent to which temporary leys should replace permanent grass in British agriculture. Ploughing the land, improving its condition, and sowing good species and strains of grass, if successfully accomplished, undoubtedly gives a superior sward to the great bulk of the indifferent permanent grassland in this country. A temporary ley can be made to conform to definite requirements by sowing a seeds mixture suitable to the conditions and the needs of the farmer, and young grass on a fertile soil will be of a greater productivity than a poor permanent sward. Modern developments in the technique of laying down land to grass, a better understanding of how to deal with the practical problems, and the availability of seeds of better strains of herbage plants, have revolutionised the outlook in connection with the use of temporary grass on the farm. The temporary ley differs from permanent grass only in that it is a means of concentrating on the better types of herbage plants grown on a soil which has been adequately prepared. It is the greatest refinement of the cultivation of grass as a crop in a system of alternate husbandry where grass occupies an accepted place in the farm rotation.

Temporary leys may be of short or long duration, the upper limit usually being in the nature of ten or twelve years. When temporary leys are left down for longer periods, they commonly lose their character as temporary leys, except under the most favourable conditions of management and environment. This results from the sown species gradually being replaced by natural colonisers, which may or may not be desirable species. In general, a temporary ley maintains its essential features in terms of the sown species only in so far as these species are suited to the conditions and are skilfully managed, and this maintenance requires a high standard of management if the sward is to be kept in full productivity. Temporary swards are essentially artificial products of an unstable nature obtained by blending highly productive herbage plants which demand careful attention for their preservation. If this attention is lacking or misdirected, the sward will deteriorate, and the whole purpose and value of the ley are lost.

The skilful use of temporary grass in all its forms gives the farmer much greater opportunities to exploit grass as a crop than if he is dependent only on permanent swards. Greater productivity by the use of superior species and strains of herbage plants grown on land which has been specially prepared and cultivated to receive them is an obvious advantage. For this purpose the most appropriate kinds of plants for the type of ley and the method of management can be chosen. On the same farm, special leys can be established to supply grazing at different times of the year, to cut for hay, or to make into silage or dried grass. The temporary ley is, in fact, the only means of engaging in the most intensive systems of grassland husbandry, which not only provide the means of developing more intensive livestock enterprises, but by a combination of both of these interdependent aspects of farming, enable a more productive agriculture to be established.

The development of the temporary ley and its incorporation into some system of alternate husbandry is one of the means by which increased agricultural production is being envisaged in many parts of the world. Rotational practice based on alternate husbandry simply implies that all the land on the farm, which it is practicable to so use, will carry temporary grass and tillage crops in an alternating system of crop sequence. Each field thereby benefits in turn by cultivations associated with tillage crops alternating with the recuperation resulting from a period under well-managed temporary grass. This country is particularly suited over wide areas to such systems of farming, the most appropriate tillage rotations and type of temporary ley varying in different localities and with the circumstances. The practicability of establishing farming systems of this type based on the temporary ley is a valuable asset to British agriculture, and the value of the grass cover for maintaining and improving soil fertility cannot be exaggerated.

Although the grasses are the obvious and dominant feature of the various types of grassland in this country, it is important to realise that cultivated and uncultivated plants of the family Papilionaceae, commonly referred to as leguminous plants, are extremely valuable components of agricultural swards. This

37

country is fortunate in the possession of leguminous herbage and forage plants which are eminently suited to blending with grasses in the formation of various kinds of swards. These leguminous plants, the most important of which are the clovers, are similar to the grasses in that there are certain types which are particularly adapted to grazing or cutting, and when mixed with grasses they play an important part in the formation of the continuous and dense cover of vegetation which characterises good agricultural grassland. The grass-clover mixture is an essential feature of the better types of permanent and temporary grassland in this country, and the high potential economic value of the best swards is dependent under many conditions on this vegetational association.

Leguminous herbage plants not only have a special value in improving the palatability and nutritive value of herbage, but their presence in a sward usually improves its productivity. The action of leguminous plants on the soil by their deeply penetrating roots and their nitrogen-fixing properties is an important consideration in relation to fertility, and one of the accepted means of improving poor grazings is by the introduction of a legume such as wild white clover. Similarly, all recognised methods of good grassland husbandry are directed towards maintaining a suitable balance of grasses and legumes in the herbage according to the particular type of sward that is desired.

The great difference between the natural occurrence in this country of grasses on the one hand, and the leguminous plants, particularly the clovers, on the other, is that whereas the grasses will grow sufficiently well, if not abundantly, under such a wide range of conditions, the clovers are more particular in their requirements. Consequently, there are large areas of permanent grassland, not to mention also of rough grazings, where leguminous plants are either virtually absent or else very poorly represented. It is only when the growing conditions are particularly suitable, as on calcareous soils, or where special efforts have been made by the agriculturist to encourage their growth, that a good representation of clovers is found.

Grasses and clovers, or other suitable leguminous plants, such as lucerne, are the important and valuable components of agricultural grassland of various types. It is seldom, however, that these are the only components, and in most cases there is a contribution in varying amounts by miscellaneous herbs. Some of these herbs have a certain value and contribute to the stock feed, while others are useless, reduce the value of the sward, and must be classed as weeds. The value of a sward has, therefore, to be based on several characters with regard to the plants that enter into its composition. In the first place, the representation of the grasses, leguminous plants, and miscellaneous plants must be considered, and then the species which compose each group must be assessed according to their economic value. Therefore, the botanical composition of grassland is the basis of its agricultural worth, and it is only in terms of the contributory plants that a true assessment can be made. Good grassland requires the use of the finest types of herbage plants, and the difference in stock-carrying capacity of poor permanent pasture compared with first-class permanent pasture is enormous, and is due to botanical composition and management. But the greatest measure of control over botanical composition, and the highest expression of productivity are offered by the temporary ley, for which the land may be specially prepared and the appropriate species and strains of herbage plants chosen for the growing conditions, the type of sward, and the method of management.

Chapter VI

GRASS SPECIES AND STRAINS

There are well over a hundred different species of grasses found in the British Isles: most of them have no agricultural value, a considerable number are weeds which must be kept under control, while a very few have any agricultural usefulness. Of the useful species, and only about twenty can be so regarded, some are valuable because they colonise naturally large areas of permanent grassland, while a smaller number constitutes the more or less cultivated forms which have to be sown and managed carefully to maintain them. If the grasslands of the country were more intensively managed, the really useful grasses would be largely confined to those in the last category, and the less valuable, but at the moment most widely occurring, species would be very much reduced in their contribution. There is a limit to the reduction in the number of species that could be effected because some are suited to growing conditions and forms of management that would not do for others. Nevertheless it would be possible to eliminate certain species from agricultural grassland, and to reduce the enormous acreage occupied by other indifferent species merely because the management is bad.

Particular species of grass have special characters and attributes which make them valuable for different agricultural conditions and purposes. It is not possible to judge the agricultural value of all species on the same criteria, although ultimately the virtues of a grass depend on its capacity to produce the largest amount of nutritious food under any particular growing conditions and farming management. But the wide range of conditions, and the specialised and varied forms of management, under which grass is grown in this country demand sometimes special assessment of agricultural worth. A species may have a special value because it thrives where no other species grows successfully, or even survives, although it may be an indifferent species under good conditions with regard to its yield and nutritive value. An otherwise indifferent species may have the virtue of being drought resistant, or be capable of withstanding very acid conditions. Of course, it is equally true that a valuable species under intensive farming conditions may be quite unsuitable for a wide range of conditions where an inferior grass holds its position simply because it persists where the other would disappear.

It may be justifiably argued that the most productive grassland can only be expected where the conditions are suitable, and that if the standard of grassland husbandry were improved, there would be no need for growing inferior species. If the economic conditions justified the expense, a great deal could be done to improve the general status of the grasslands of this country, but there would still be a reasonable proportion of the grassland which was not suited to intensive management, and which would continue to be dominated by species which are regarded as inferior under better conditions. With the present wide range of grassland types, it is necessary to consider the various species according to their contribution and their suitability to fulfil some useful agricultural purpose in the many conditions and under the particular forms of management that may be experienced.

From the agricultural point of view one of the most fundamental characters that distinguishes grass species is their suitability for soil types and particularly different levels of fertility. Some species are regarded as low-fertility types; they colonise naturally the various types of indifferent permanent grassland and flourish when grassland is neglected. Other species are found only under good natural conditions and with intensive management of the better permanent grazings and rotational grassland. They are the high-fertility types in that they respond to good conditions and are the highest-yielding grasses with good management. This broad distinction is not only necessary for understanding the main types of agricultural grasses, but it affects also their position and status as cultivated plants. The high-fertility grasses are, for the most part, those species that have been taken into intensive cultivation because of their responsiveness to good treatment. They are suited to various types of intensive husbandry such as rotational grass and ley farming, and they have proved amenable to improvement under cultivation and by plant breeding.

41

It is largely within the rotational and high-fertility grasses that strains or varieties have been developed by various means. These strains are in many ways akin to the varieties in other crops, and they have had very important effects on grassland husbandry in recent years. In some of these grasses several strains are now available, each possessing very distinct agricultural and botanical characters, and it is important for the grower to understand the general significance of this position and the possibilities of being able to choose different types of any one grass species.

When the agricultural value of particular grasses was first recognised it became the practice to collect seed for sowing on arable land. In time, a trade sprang up in grass seeds and fields were set aside specifically for the purpose of growing seed, and the propagation of the more widely grown species such as perennial rye grass. As this trade developed it became customary for this country to import seed of various agricultural grasses, and the seeds houses distributed them to the growers for sowing down fields to grass. In this way the position gradually arose that much of the temporary grassland in this country was being sown with seed which may have been grown in various European countries, in the U.S.A. or Canada, or in different parts of Great Britain, and it was possible to obtain the same species grown for seed in different parts of the world and handled by the seed trade as commercial seed products.

Eventually it became obvious that consistently growing a grass for seed year after year might be working for the benefit of the seed grower, but it was not necessarily the best thing for the farmer who was to use the grass for its herbage and not for its seed. It was equally obvious that stocks of seed of one species that had been grown in different countries under entirely different conditions were quite distinct when their behaviour was observed under farming conditions in this country. This is explained by the fact that these grass species are cross-fertilised and they do not breed true to one uniform type. Thus when seed is taken consistently year after year, the types which will be predominant will be those that run up to seed most readily and produce the greatest amounts of seed under the conditions of seed growing. Such types may or may not be the most suitable for the farmer who includes them in a seeds mixture for a grazing or hay sward under entirely different conditions from those under which the seed was grown.

It was from arguments and knowledge of this kind that the importance of strain in grass species began to be appreciated, and it was seriously questioned whether the best use was being made of the great range of herbage plants that existed in nature. Instead of importing seed of grasses which had been grown in foreign countries, would it not be more reasonable to use seed from plants which were growing in good permanent pastures in this country? Such plants had shown themselves to be adapted to the conditions under which they were growing: they were native strains which by the action of natural and agricultural selection had proved themselves. In this way the idea of the indigenous or native strain came into being, and the realisation that nationality in herbage plants was a matter for serious consideration stimulated a new interest in the grasses which were being used for sowing agricultural swards of all types. Indigenous strains of the more important grass species were developed and the relative value of such strains compared with the commercial strains became a matter of much controversy and considerable investigation. In various parts of this country local indigenous strains characteristic of regions or counties were offered to growers, and considerable reputations were built up for some of these strains.

The intensive study of these grasses which resulted from these developments rapidly led to the realisation of the extreme variability that was present in many of the most valuable grass species such as perennial rye grass, cocksfoot and timothy. It was then but a natural step to consider how far these species could be further improved by controlled and planned breeding to suit the various purposes to which they were put agriculturally. These important species were examined exhaustively by taking individual and separate plants of different types and from various localities. The individual plants were tested by dividing them up and obtaining as large a number as possible of vegetative progeny so as to be better able to judge their value and suitability for special purposes such as grazing or cutting for hay. In this way, by selecting suitable types, pedigree or bred strains were developed by the judicious mixing of lines of known origin and proved ability, conforming to the desired type which the plant breeder was endeavouring to produce.

The position has now been reached when the farmer can choose between several distinct strains in the important grass species, and also in some of the clovers on which similar work has been done. The main distinction is between the commercial, the local indigenous and the pedigree or bred strains. Commercial strains may be of foreign or native origin and their characteristics and agricultural use depend very much on where the seed was grown. Local indigenous strains are the product of natural selection in a restricted locality, and in the nature of the case are characterised generally by being long-lived pasture types. Pedigree or bred strains may be pasture or hay types, but the original plants which were used to make up these strains are usually native, or indigenous, and the product is commonly a leafy indigenous type capable of a high degree of persistence in a sward. The variability between different commercial strains is in some cases very great, but as a generalisation it may be said that these strains tend to be of the more erect hay type, with early growth and lacking in persistence. But in all strains the important points to consider in judging their suitability for any purpose are their origin, history and method of production. A pedigree strain is not necessarily a superior strain for all purposes; there are good and bad pedigrees, and also pedigrees which suit a strain to one set of conditions or management, but may make it totally unfit for other conditions and other forms of utilisation.

The important grass species which contribute to the grasslands of the agriculture of this country may then be divided broadly into two groups. First, there are those species which are included in seeds mixtures of various kinds, and which are the most valuable grasses for rotational swards and the most productive permanent grasslands under intensive management. Secondly, there are species which, though not usually sown in seeds mixtures, colonise naturally to a greater or less degree and contribute under some conditions to a large part of the permanent grasslands. The grasses of the first group are few in number and are essentially the cultivated species. They occupy a relatively small proportion of the total grassland acreage, but they are the most productive and need first consideration.

Of all the cultivated grasses, perennial rye grass (*Lolium perenne*) has the longest agricultural history and the widest use in this country. It is found throughout Britain on many types of grassland and it also occurs more or less as a weed and a natural colonist of waste places. When it was first recognised as a valuable agricultural grass it was called 'ray grass', and its corrupted name to-day does not mean that it has any relationship with the cereal of the same name. Rye grass is outstanding as a pasture plant because it is a vigorous grower, producing high yields of nutritious growth on good soils, forming a close turf when well managed. It is an easy grass to grow from seed and rapidly establishes itself where conditions are suitable, but like all good agricultural grasses, it will not persist under extreme conditions. Owing to its capacity for quick establishment from seed it can be used for short leys, while its great powers of persistence make it equally valuable for permanent swards and long leys. There is no other grass that has such wide applications, and perennial rye grass is usually the basis of most seeds mixtures.

As befits such a valuable and widely grown species, perennial rye grass is available in many strains. This has added to the usefulness of the grass because appropriate strains may be chosen for particular purposes. There are foreign commercial strains; Irish, Scottish and English commercial strains; local indigenous strains; and pedigree strains developed in this country and abroad. Some of the commercial foreign strains are short-lived, stemmy types, and are not suitable for growth in this country; others are hay types which can be used for leys of one or two years. The British and Irish commercial stocks are on the whole superior to the foreign, being better suited to the conditions of this country, but they also tend to be more useful for the shorter leys from which hay is to be taken. There is considerable variability between some of the local indigenous strains, but they are essentially suited to long-term and permanent grazing swards, because the seed is only harvested from old swards. The pedigree strains developed in this country may be of grazing, hay or dual-purpose type, an example of each type being S 23, S 101 and S 24 respectively. These bred strains are all leafy, long-lived forms and can usually be relied upon for more sustained growth through the season, although some are rather late starting growth in the spring compared with the commercial strains. Most of the foreign pedigree strains, although they show considerable variability in behaviour, tend to be intermediate between the indigenous and British commercial stocks when grown in this country.

Although perennial rye grass can be the principal ingredient of most seeds mixtures, and therefore of most lowland agricultural swards, it is clear that the choice of strain is really important. Persistency, habit of growth, season of growth and leafiness are all strain characters, and to obtain the best results it may be necessary to blend two or more appropriate strains to obtain the kind of sward that is wanted for the particular conditions and management. Differences in persistency and leafiness between the various strains require especial attention, because it is in these characters that strains can vary so markedly, and which are so important agriculturally.

The other commonly used rye grass is Italian rye grass (*Lolium italicum* or *multiflorum*) which is regarded botanically as a distinct species from perennial rye grass, although there are hybrid forms occurring naturally and as the result of artificial hybridisation. Italian rye grass is a biennial, and is a very valuable grass for one- or two-year leys, because of its very rapid leafy growth from seeding. It can be used for grazing or for hay. It is particularly valuable for autumn grazing in the seeding year, and there is no other grass that quite equals it for the amount of early growth it makes in the following spring. The great vigour of growth from seeding that characterises Italian rye grass has given it a special value as a 'nurse' for other grasses when establishing a ley from seed, but this character also makes it very liable to smother out slower-growing species if sown too thickly and allowed to grow without any check. Italian rye grass is essentially a 'cultivated' grass, and was introduced into this country over a hundred years ago. There are various stocks of this grass available, but most of the seed used in this country is grown in Northern Ireland and very little has been done to develop strains for specific purposes or conditions. There is, however, a very quick-growing annual rye grass called Westernwolths grass, which is sometimes regarded as a variety of Italian rye grass, although it is more usually given the rank of a distinct species—*Lolium woldicum.* This grass was introduced from Holland in the early part of the present century, but is very little used in the agriculture of this country.

Perennial rye grass is the great standby for temporary and permanent grassland of high productivity in this country, but there are many conditions where this grass is not at its best. On infertile soils where drought is a danger, cocksfoot (*Dactylis glomerata*) is a better grass, and it should totally or partially replace perennial rye grass under such conditions. Being a large and vigorously growing perennial grass, cocksfoot is suitable for temporary and permanent swards, and it does well when mixed with other similar types of strong-growing grass. Its great virtues are its hardiness and ability to thrive where perennial rye grass fails, combined with heavy yield as hay or grazing, which is associated with its sustained growth and power of recovery. Although cocksfoot can do better than most grasses under poor conditions, like all grasses suited to intensive husbandry it is at its best on good soils with ample water supply.

Cocksfoot has not such a long history as a cultivated grass as has perennial rye grass, but it has received as much attention from plant breeders in this country. There are foreign and indigenous commercial strains, some coarse and stemmy and not persistent, and others good for hay or early grazing. The pedigree or bred strains show a great range of types varying from the extreme pasture S 143, to hay types like S 37, and among the indigenous bred strains there are types suited to different growing conditions. Cocksfoot has the reputation for being an unpalatable grass, and although strains appear to differ in this character, the species as a whole needs careful management to prevent it becoming too old before use as grazing or hay.

Between them, perennial rye grass and cocksfoot can fulfil the requirements for a wide range of conditions in this country, but there are certain purposes for which timothy, or catstail (*Phleum pratense*) is specially suited. Timothy's particular use is on the wetter soils, either heavy or peaty, and it stands low temperatures and conditions at high elevations, but is not a good drought resister. There are, however, a large number of forms of the grass, and different strains have been developed varying from prostrate grazing types for upland conditions to erect leafy kinds adapted to the lowlands. As in the case of perennial rye grass and cocksfoot, more persistent pedigree strains of leafy habit have been bred, as distinct from the stemmier and shorter-lived commercial forms. One of these bred strains, S 50, is an extreme pasture type and creeps by means of horizontal stems, while S 48 is suitable for permanent grazing swards in the lowlands, and S 51 is a leafy hay type capable of producing heavy aftermath growth.

Rotational mixtures are usually based on perennial rye grass, cocksfoot and timothy, and the same species can be used for swards of a more permanent kind. But on fertile and moist soils, meadow fescue (*Festuca pratensis*) can be used to replace cocksfoot and perennial rye grass, and under these conditions it mixes well with timothy. Meadow fescue may be used for hay or grazing, and where the conditions are good, it gives heavy yields of palatable and nutritious growth for leys of three years and over. Some of the foreign commercial strains are poor in this country, and there are better grazing and hay strains that have been developed here on the basis of leafiness, as, for example, the strain S 53. The closely related tall fescue (*F. elatior*) also has its uses as a very heavy yielder of rather coarse herbage for grazing or hay on wet and heavy soils, but this grass is not used extensively in Britain. It has to be managed carefully to prevent it becoming too coarse, but it has considerable value under extreme conditions of wetness and also on some classes of poor hill land.

The grasses so far mentioned include all those of more robust

48

growth which can be relied upon to give the highest production of forage of any of the more generally cultivated grasses in this country. There are other species which can be used for sowing long-term grazing leys, and although not so productive are valuable for certain purposes and conditions of growth. Of these, rough-stalked meadow grass (*Poa trivialis*) is in many ways outstanding as an excellent creeping grass for grazing purposes on moist, fertile soils, while the related smooth-stalked meadow grass (*Poa pratensis*), which has underground creeping stems, is suitable for grazing on dry soils and where the rainfall is low. These two grasses are rather slow establishing themselves from seed, and are only useful for long leys or permanent grass, in which they may be very valuable for forming a close 'sole' to the pasture. In exposed conditions, and on dry soils, crested dogs-tail (*Cynosurus cristatus*) can be used for similar purposes, but this grass is not a creeper and tends to produce unpalatable inflorescence-stems which stock will not graze, but which are this species' means of persisting and multiplying.

There are few other grasses that are used agriculturally by including them in seeds mixtures, except under exceptional conditions or for special purposes. There are, however, several species that colonise naturally in various types of grassland, some of which are widely distributed and contribute extensively to permanent grazings throughout the country, while others are more localised. The best known and most widely distributed of these are the various forms of *Agrostis*, generally called bent grass, which have already been referred to as important plants in many of the grazings in Great Britain. The value of *Agrostis* depends on the particular type, and the conditions under which it occurs, some of the loosely creeping types being very inferior, while the more compact leafy forms supply useful grazing where other grasses will not persist. Under good grassland conditions, *Agrostis* in all its forms is undesirable and should not be allowed to occupy space that would be better devoted to more valuable species. Indeed, under these circumstances, *Agrostis* may be regarded as a sign of grassland neglect and deterioration, and the sward should be ploughed.

On dry soils, and more particularly those of a chalky nature in open and exposed situations, useful fine-leaved *Agrostis* forms are found associated with other fine-leaved grasses. The best of these are the fine fescues—sheep's fescue (*Festuca ovina*), hard fescue (*F. duriuscula*) and red fescue (*F. rubra*)—all of which give valuable sheep grazing and are the only species that persist satisfactorily under these conditions and management. The golden oat grass (*Trisetum flavescens*) is also commonly found on the chalk soils with the fine fescues, and these species are sometimes used in seeds mixtures for sowing such areas to sheep grazing swards.

Sweet vernal grass (*Anthoxanthum odoratum*) is a species that has a reputation among farmers as a useful constituent of meadows and pastures. Its virtues have been overrated largely because it contributes, out of all proportion to its value, to the sweet smell of hay, but its actual productivity is low and its palatability very inferior. Sweet vernal has one virtue, and that is earliness of spring growth, a character that it shares with meadow foxtail (*Alopecurus pratensis*). These two grasses are the earliest species in agricultural swards, but neither can be recommended for sowing in seeds mixtures; sweet vernal because of its other inferior characters, and meadow foxtail because of the difficulty of establishing it from its very expensive seed, although it is a valuable grass and worth encouraging.

Where extreme drought resistance is required, there are three species available which are capable of higher productivity than any other grasses in general agricultural use. Tall oat grass (*Arrhenatherum avenaceum*) has greatest use as a hay plant for short leys, Schrader's brome grass (*Bromus schraderi*) has a similar application, while awnless brome grass (*B. inermis*) is better suited to long-duration swards because of its underground creeping habit. None of these grasses should normally be sown by themselves, and are much better grown in mixtures, particularly with a leguminous plant like lucerne or sainfoin. The brome grasses are seldom used in this country, and tall oat grass, although well known as a widely distributed hedgerow plant in Britain, has very limited value for agricultural purposes.

There are other grass species which contribute to the agricultural grasslands of this country, but they are found mostly in limited amounts and on rough grazings. They are never included in seeds mixtures and they do not even colonise the better types of grassland to contribute anything of value. They do not exist as a result of any controlled agricultural management, and they cannot be regarded as in any way cultivated plants, deliberately encouraged or tended by the farmer under good grassland conditions.

The grass species used in agriculture are all cross-pollinated by wind. This necessitates the use of a special technique in breeding, and introduces certain complications in the building up of bred strains, the maintenance of 'pure' stocks, and the multiplication of these stocks in seed production. In grass breeding it may be said in general that the breeder does not rely on hybridisation to create new variability and types, but he exploits the natural variability already in existence by selecting the most suitable types for strain building. But having attained the desired type in the form of a nucleus stock of a pedigree or bred strain, the next stage is to multiply this stock for use in agriculture.

The dangers of natural cross-fertilisation with other strains of the same species, or even with plants flowering in hedgerows, or on agricultural or waste land, have to be guarded against by taking adequate precautions. Suitable isolation of breeding material, nucleus stocks and large-scale seed-multiplication fields has to be practised, and the organisation of the production of recognised strains needs careful and constant supervision if strains are to be maintained. Further complications are met with in the commercial handling of herbage-plant seeds, because it is normally impossible to distinguish between different strains on their seed characters, while the necessity of constantly resorting to the original breeding stocks to replenish the seed supply is an obvious procedure. Therefore, although the development of improved strains has been the most significant advance in grassland husbandry during the past thirty years, this great advance has brought with it the necessity of developing and maintaining an adequate organisation to deal with the situation.

Chapter VII

THE CEREALS

The cereals belong to the same botanical family—the Gramineae —as do the grasses, but whereas the grasses are grown agriculturally for their vegetative parts for feeding to livestock, the cereals are cultivated for their grain which is pre-eminently a human food, but is also of great importance in maintaining intensive livestock industries. As far as is known, all the important civilisations of the world have been dependent for their establishment and development on the cultivation of one or more cereal crops whose grain has supplied the staple food of the human population. With the domestication of livestock, the cereals soon came to occupy an important position in contributing to their maintenance by the use of the grain and the straw. This latter use developed a secondary but very important function in the economy of intensive agriculture by which the cereal straw became the basis of the method of returning plant and animal residues to the soil through the making of farmyard manure.

Some of the cereals, such as wheat and barley, appear to be the most ancient of the cultivated agricultural plants at present in existence, and the vast increase in the human populations of the world during historical times has been made possible by the successful extended cultivation of these grain crops for human and animal food. The cereal production of the world, particularly in temperate regions, has increased enormously in recent times principally by man's colonisation of new areas which have proved suitable for the cultivation of grain crops, but there has been in addition more intensive cultivation in the older centres of human activity.

Although it seems obvious that the cereals were derived originally from wild grass-like types of the Gramineae, the evolution of these plants under domestication has reached such an advanced stage that it is difficult to trace with exactness their botanical history. The tremendous botanical variation in some of the cultivated cereals shows the effects of isolation by man of a wealth of forms which has no parallel in other crops and is indicative of a great capacity for adaptation. In some of the cereals there are wild grasses of very close botanical relationship; in others there are primitive wild-grain types which are weed-like, but this does not necessarily mean that the cultivated cereals were actually derived by selecting from these wild grasses or primitive forms, and all the evidence indicates that the so-called 'wild progenitors' of the cultivated cereals no longer exist in their original state.

The great value of the cereals is due to their capacity for producing a large amount of 'seed', which is not only edible, but is of a highly nutritious nature. The 'seed', or grain, is really a fruit, rather akin to a nut, but with the outer covering forming a thin skin instead of a hard shell. The true seed inside is composed mostly of a food reserve of a concentrated nature, consisting largely of starch, but with smaller amounts of protein and oil. The whole grain is, therefore, a very convenient source of food which can be eaten whole after cooking, as in rice, or which may be ground and cooked or baked in various ways as in wheat, rye, oats, barley and maize. A further very valuable attribute of the cereals is that the grain may be stored to carry over from one harvest to the next, while the ground flour or meal can also be preserved for future use. All these characters mean that not only can cereals be used as the principal food locally in the area of their cultivation, but they may be transported over great distances with comparatively little expense due to special storage and transport equipment.

The almost universal success of the cereals has been made possible by the great adaptability as cultivated plants of the more important ones, and by the wide range of forms in this group of plants as a whole. This has allowed man to develop some cereal or other in practically every climate which he himself finds congenial or bearable. Cereals also lend themselves very well to tillage operations and are easy to manage as farm crops. All the important cereals are annual plants which may be sown and harvested within one growing season, or, with some varieties and in certain climates, they may be sown in the autumn, allowed to stand the winter, and harvested the following year. Cereals are amongst the easiest crops to handle in the field when it comes to harvest time, and from primitive methods of hand labour there has been a gradual development of increasing mechanisation to the present day, when grain may be cut, threshed, dried and put into bags by machinery. In modern systems of rotational and mixed farming, cereals fulfil a vital role in a balanced agriculture not only as supplying a necessary part of the diet of human beings and stock, but also in contributing the straw for the manufacture of farmyard manure. In some parts of the world, however, a cereal is grown as the only crop, either extensively as with wheat in the prairie provinces of Canada, or intensively as with rice in the monsoon areas of Asia. These two examples show cereal cultivation at two opposite extremes with regard to labour; in Canada, mechanisation and low labour in terms of manpower are the characteristic features, while in Monsoon Asia rice is the only cereal which can support the dense population under the system of intensive cultivation of the arable land which is economically necessary.

The dominating position of the cereal crops in the agriculture of the world and in the life of the people may be realised from man's dependence on cultivated grains for his food. These grains are the principal source of food energy and the cheapest form of carbohydrate calories, and the greater part of the world's population depends on cereal grain of some form for the bulk of its food energy, while wheat and rice together are easily the most important foods in the sustenance of human beings. Wheat is grown widely throughout the temperate settled agricultural areas, where it is in general the most important cereal, but rye is more important in some countries, and barley may share a position of equal importance with wheat in others. Oats, except in very restricted areas, are not grown as a staple food for human beings in any country, but owe their economic position in agriculture to their value as a stock feed. In the wetter and warmer parts of the world, rice takes the place of wheat as the staple cereal for human consumption, while maize, millets and sorghums occupy a similar position in certain subtropical and tropical countries where the temperate cereals cannot be grown.

The vast world consumption of cereals and their products by human beings is largely a reflection of the low standard of living and dietary which characterise the great proportion of the population. A study of the dietaries in different countries, or the dietary of people of different income levels, shows that the proportion of cereals consumed increases as the standard of living declines. In certain rice-eating countries, very little else is consumed by the people, and the standard of living may in some circumstances be little above subsistence level. In contrast to this, where the standard of living is high in certain countries which consume wheat as the principal grain, less than one-third of the people's diet is composed of cereal products.

In the British Isles the cultivation of cereals has ceased to be the dominating feature of the national agriculture. The standard of living, which by comparison with many other countries is high, has led to a relatively varied diet with a consequent considerable reduction in the amount of cereal products eaten by each person. This, of course, is true only in normal times of peaceful world trade, but in wartime, and the aftermath of war, cereal production and consumption in this country show considerable increases. The same considerations also hold for the balance of cereal production in Britain, where the emphasis is normally not on the cultivation of staple foods for human consumption, but rather on the production of cereal products which form an essential part of the mixed farming and stock husbandry which characterise the country's agriculture. It is true that the greater proportion of the wheat crop is milled for flour, and over half the barley crop is brewed for beer, but neither of these methods of utilisation is responsible for maintaining the people. Further, the oat crop, which occupies the greatest acreage of the three cereals in Great Britain is used almost exclusively as a livestock feed, the straw being of great agricultural value for this purpose.

The relative cheapness of cereal cultivation and production, particularly in a system of mixed farming in which other crops are grown in rotation, more or less assures an important place for grain crops under arable farming conditions where the climate and soil are suitable. In Britain, the distribution of the cereal crops and the dominant position of either wheat, barley or oats in various parts of the country, is decided principally by the climate and whether the crop is grown for marketing the grain or feeding to livestock. Soil conditions as a whole do not limit cereal cultivation in this country because the seed is very accommodating, and can be sown successfully where many other crops would fail. The important factor that does limit cereal cultivation, particularly from the point of view of grain production for sale, is the climate for ripening and harvest. When judged by the standards of the British climate, cereals are not heavy in their moisture demands for the production of high yields of grain, but they do require reasonable sunshine and warmth for ripening satisfactory crops of good grain quality. It is, therefore, understandable that the important grain-growing areas in Britain are found where the rainfall is low and the summers have most sunshine.

But the whole position of cereal production in this country is modified by the importance of cereals as livestock food. Of the three cereals cultivated, wheat is least used for this purpose, and the distribution of the wheat crop is therefore decided on the possibilities of its cultivation as saleable grain. Similar considerations hold very largely for barley, which is grown principally for the maltster and brewer, but with this crop there is the alternative of economic cultivation solely for feeding to livestock. This is done either on soils too rich for malting barley, or more generally on thin soils in drier districts where oats are not successful, although there are small and isolated areas of barley cultivation for stock in the most northern parts of Britain where the season is too short to grow any other cereal. But it is the cultivation of oats which shows the greatest association with the livestock industry of the country because this crop is grown almost entirely for feeding to farm animals. Because of the importance of the straw as fodder, and the suitability of the crop to the moister and cooler parts of the country, oats can be grown economically where the other two cereals cannot be considered. Oats can be cut and used successfully in an under-ripe condition, and there are special varieties suited to this purpose and to climates that are unsuited to intensive grain production. This extends considerably the range of economic cereal production, and although the areas of intensive grain production in oats are the districts of good arable farming, the extent of the cereal acreage is very largely dominated by the wide distribution of the oat crop as a grain and straw producer for livestock.

Cereal cultivation in Britain consequently shows many aspects of the agricultural importance, nutritional value and economic significance of these cultivated plants. The value of the cereal as a rotational crop, easy of cultivation, adaptable to the growing conditions, accommodating in soil requirements and amenable to different kinds of management is well exemplified. Each of the three cereals cultivated is an example of the virtual monopoly of the cereals as intensive producers of concentrated food by virtue of their high grain production—human food by wheat, and animal food by barley and oats—with the important secondary product of fodder from the straw. Finally, there is the special virtue of these crops in providing products of particular market value as in the bread, biscuit and confectionery qualities of wheat, the malting and brewing quality of barley, and the oatmeal quality of oats. The cereal grain has no substitute as an easily produced, conveniently handled and economically transportable concentrated food which can be stored indefinitely and used at any season of the year as the basis of the diet of man and livestock.

Chapter VIII

WHEAT

Cultivated wheat is undoubtedly one of the most ancient of agricultural crops in temperate countries, and the oldest civilisations of North Africa, Europe and Asia Minor knew wheat as a bread corn. From its ancient centres of cultivation, wheat has spread to every temperate country where there is a settled agriculture, and it is even grown in tropical and subtropical countries at high elevations. During its long history as a cultivated plant many thousands of distinct forms of wheat have arisen, until at the present time there is a bewildering number of types cultivated throughout the world. There is no other cereal that has achieved such a universal importance as wheat, and although it is not the staple food of as many people as is rice, it is produced in as great a quantity, and occupies a far greater acreage in the world than any other tillage crop. During the last 100 years, wheat cultivation has spread enormously throughout the world until to-day there are several very large centres of wheat production in the temperate zones, with Europe, the U.S.S.R. and North America contributing the great bulk of the crop. With the increase in the population of the older centres of civilisation, as in Europe, and with a greater demand for wheat as a bread corn in preference to other cereals, a tremendous international trade has been developed in wheat as new areas have been opened up for cultivation. The world wheat crop is now a matter of international concern, and crop failures in any of the large wheat-producing areas may have serious consequences for the big wheat-importing countries.

Wheat has been cultivated in Great Britain since the New Stone Age, and the successive invasions which this country suffered up to the Norman Conquest were in all probability responsible for the introduction of new forms and varieties. Even before the Christian era, wheat appears to have been a well-established crop in the settled parts of southern England, and during the Roman occupation wheat cultivation undoubtedly spread. But in these early historical times barley was also grown as a bread corn, and later rye and oats were also used to such good purpose that up to the beginning of the eighteenth century the majority of people were eating bread made from cereals other than wheat, because the wheaten bread was so expensive. It was not until the first quarter of the nineteenth century that wheaten bread became virtually the only type of bread eaten in this country, but by this time the population had so increased that the home-grown wheat crop could not meet the demands of the consumers.

Great Britain has now become one of the largest wheat-importing countries of the world, and something like three-quarters of the requirements are obtained from overseas. With the country's population at its present level it is impossible to grow sufficient wheat at home, and with the competition from imported wheat, better suited to bulk handling by the miller and baker, and giving a type of loaf more popular with the public and at a lower cost, home-grown wheat production in peace time has become of minor importance in supplying the daily bread of the British people. Although something like two-thirds of the home-grown wheat crop is milled, only a small proportion of this flour is normally used for bread-making by blending with imported wheat. Much of the flour is used for biscuits and cake flour, a purpose to which the bulk is better suited, while the remainder of the crop is fed to poultry and other stock. The result of this economic position has been that the wheat acreage in this country has contracted considerably below the amount that would be grown if world wheat prices were higher and if there was a strong demand for home-grown wheat for bread-making purposes.

The cultivated wheats are all included in the botanical genus *Triticum*, of which there are several distinct species grown in various parts of the world. These species differ in their economic importance, their suitability for making various kinds of articles of food, and their adaptation to particular growing conditions, but by far the most important is *Triticum vulgare*, the species which is commonly called 'the bread-wheat species'. There are many hundreds, even thousands, of different kinds or varieties of bread wheat in existence, and the species is usually regarded as being the most advanced from the evolutionary point of view of all the cultivated kinds. The great world development and expansion of wheat cultivation has been due to the successful adaptation and usefulness of the bread wheats, combined with the breeding of improved varieties for special conditions and markets.

Although no other wheat species is as suitable for bread-making as forms of *Triticum vulgare*, and few of the other species are as successful from the point of view of yielding capacity and adaptability to intensive and extensive cultivation under a wide range of conditions, there are certain parts of the world where other species are grown exclusively or in addition to the bread wheats. The most important of these other species are the Hard or Macaroni wheats (*T. durum*), the Club wheats (*T. compactum*), the Rivet or Cone wheats (*T. turgidum*) and the Spelt wheats (*T. spelta*). Each of these has a suitability for the production of a certain class of wheat, and an adaptation to particular growing conditions, but none has such a general economic importance as the bread wheats.

In this country, with the exception of a very small acreage of Rivet wheat, the cultivation of wheat is devoted to varieties of bread wheat which are the most suitable types for the growing conditions and the methods of cultivation. The crop is particularly suited to the lower rainfall areas, and on the more fertile soils yields of grain can be obtained which are only equalled in a few intensively farmed and restricted areas in other countries. A suitable climate, freedom from uncontrollable epidemic diseases, and rotational farming of an intensive type, make the wheat-growing areas of Britain admirably adapted to this crop from the point of view of grain production.

In Britain, wheat is typically an autumn-sown crop; it occupies the greatest acreage of the total autumn-sown cereals, and its production in this country means essentially the cultivation of 'winter wheat' which occupies the land for something like eight or nine months of the year. The length of the growing season varies considerably, however, because the crop may be sown at any time between late September and the end of December, although October and November are normally the best months. A certain amount of winter wheat is drilled in January and February, but the conditions need to be very favourable for this practice, and October is usually the best time.

There are few soils totally unsuited to the growth of wheat, but the economic cultivation of the crop is seen at its best on the heavier fertile loams, although satisfactory crops can be grown on clays and light to medium loams. The lightest soils, particularly if acid, are quite unsuitable, and the really wet, cold conditions of the heavy clays in the winter can only be withstood satisfactorily by some varieties. Intensive wheat production, as practised in this country, needs a good soil moisture supply with low rainfall and warm, sunny conditions in the summer for proper ripening and successful harvesting of a good sample of grain. But heavy yields can only be obtained on fertile soils; and with the strong strawed, prolific varieties now available, wheat responds well to high farming, and it is usual to take the crop in the rotation after suitable preparatory crops. In many good wheat-growing districts, clover, a seeds ley, or other fertility building crop such as field beans, peas or trefoil are common forerunners, while well-manured root crops make excellent preparation possible. Under all conditions, the growers' object is to ensure a really fertile and clean soil without running too great a risk of causing the wheat to lodge.

Wheat is the world's most important bread-making cereal because the grain has certain characters which enable loaves to be made that are more palatable and digestible than those made from other cereals. It is, indeed, impossible to make the type of loaf demanded by modern tastes from other cereals, and wheat will continue to hold its pre-eminent position as a human food simply on this character. Other cereals have similar gross chemical constitutions to wheat, but wheat alone possesses the particular protein (gluten) characters which enable the baker to produce the light-textured and otherwise attractive loaf at present in fashion.

Baking quality wheat should possess good milling characters as well as good baking characters. The grain should be free from weeds, dirt and other foreign matter, and it is necessary that there should be no weed seeds which are liable to cause tainting of the flour. Freedom from all forms of grain damage, including mechanical, heating or weathering is of importance, and the grain must be well-filled, dry and without any musty smell. A good milling wheat must be suited to the modern methods of roller milling which depend for their efficiency on a ready and easy separation of the flour from the bran, and the final flour product should fulfil the requirements of good texture and colour. Many of these requirements are dependent on growing, harvesting, threshing and storing conditions on the farm, but some are also varietal characters, inherited from generation to generation, but liable to some modification according to growing conditions.

The present position occupied by wheat growing in British agriculture depends on the extent to which the varieties which can be grown, the methods of handling by the farmer, and the growing conditions can supply a product which is in demand on the home market. Extraneous defects such as mechanical damage, dirtiness, contamination by weed seeds, mustiness, weathering and heating can all be avoided, except in most unfavourable circumstances, by good management on the part of the grower. The primary consideration is to what extent, given reasonable conditions, the farmer is in the position to market wheat which possesses the desirable intrinsic characters for its economic utilisation by the various processing methods by which it reaches the consumer.

There are many reasons why wheat growing in this country has ceased to hold the important position that it did formerly. Why, it may be asked, is home-grown wheat now for the most part regarded as inferior for bread-making when a hundred years or more ago virtually all of it was used for this purpose? The answer is that the bread-making and milling qualities of all varieties of bread wheats are not the same, and they are also affected to some extent by the conditions under which the wheat is grown. When wheat began to be imported into this country in large amounts during the nineteenth century, it was found that it made a different kind of bread which the public soon came to prefer to the bread from home-grown wheat. The millers also found that they could handle the large and uniform imported consignments more economically than the smaller and more variable lots of home-grown wheat, and the bakers soon realised, among other things, that they could make more loaves from a given amount of flour from imported wheat. But the final blow to home-grown wheat production was that the great wheat-exporting countries were able to market their wheat in this country more cheaply than the British farmer was able to do.

It was thought for many years that the climate of this country prevented the growing of the best baking wheat, and this led to an intensive study of the characters and conditions which determine baking quality. It was soon found that the protein (gluten) in the food reserves or endosperm of the wheat grain was the important character which was largely responsible for bread-making quality, and that the gluten of the best imported wheats was different from that of the home-grown. This gluten character was shown to be inherited, and different varieties possessed distinct types of gluten which made them good, bad or indifferent baking wheats. It has since been found that other grain characters affect baking quality, but although the exact explanation in terms of physics and chemistry of differences in glutens has not been fully discovered, it is still true that gluten is the important character and that variety is more important than growing conditions.

The obvious way out of this difficulty appeared to be that this country should grow the good baking wheats that were being imported, but this proved quite uneconomic because their yields are far below those of the varieties commonly grown in the British Isles. A further difficulty with regard to the growing of baking wheats in this country is that the humid conditions, even in the lower rainfall areas where wheat is most successful, lead to a higher moisture content in the grain than is the case with grain imported from Canada which now supplies most of our wheat. This damper condition of home-grown wheat leads to difficulties in storing, while millers and bakers find certain disadvantages in handling British wheats as compared with the best imported product.

Although the handicap of the high moisture content grain can only be overcome by artificial drying, a process which adds to the cost of production, it has now proved possible to offer growers in this country wheat varieties which combine high yielding powers with good milling and baking qualities. The first of these varieties was distributed in 1916 under the name of The Yeoman, and it had been produced by hybridising the high yielding, poor baking quality English wheat, Browick, with the good baking quality Canadian wheat, Red Fife. Subsequently, another selection from the same cross was released to growers as Yeoman II, and these two Yeoman wheats proved that it was possible to grow grain of good baking quality from high-yielding varieties grown in this country. In 1937 a new hybrid variety was put on the market. This variety, Holdfast, was selected from the offspring of a cross between Yeoman and White Fife, and it shows excellent milling and baking qualities combined with a capacity for giving very high yields on fertile soils and land in high condition. Both Yeoman and Holdfast are white chaffed, beardless varieties, but the former has red grain and the latter white. Neither is suited to poor soils or poor farming conditions generally, and they should only be chosen for cultivation when high yields can be expected on medium and heavy soils.

Two other high-quality milling and baking wheats have been produced by hybridisation with Yeoman as one of the parents. Redman, a red-chaffed and red-grained wheat from the cross Yeoman and Squareheads Master, appeared in 1934, and Warden, which has a white chaff and grain, was obtained by hybridising Benefactress and Yeoman and was first made available in 1938. Both of these varieties, like all the commonly cultivated varieties in this country, are beardless, and they are again suited to medium and heavy soils in good condition, although Warden is not quite as suitable for the highest farming conditions.

Some of the most extensively grown wheats in this country have no pretensions to baking quality but are particularly suited for making biscuit flours. One of the best known and widely cultivated of this class is the old variety Squareheads Master which has red grain and chaff. The origin of this wheat is not known, but it has been one of the most popular wheats in this country at any rate since the end of last century. There are several forms of this variety, and some other varieties with different names are indistinguishable from Squareheads Master forms, but taken together this group of varieties is probably still occupying the greatest acreage of wheat in Britain. Squareheads Master is a very adaptable variety and a reliable yielder on medium soils, but it should not be grown on rich soils because it will become laid. Recently a selection, No. 13/4, has been put on the market and this is capable of higher yields.

The other important group of biscuit wheats consists of certain varieties with white grain and chaff. These include the Dutch wheats, Juliana and Wilhelmina; Wilma which is a selection from the latter variety; and the English hybrid variety Victor which was obtained by crossing Squarehead with Red King and then subsequently hybridising with Talavera. In view of the suitability of Wilhelmina and Juliana for conditions in this country it is interesting that Wilhelmina is a hybrid from the old English Squarehead, with a Dutch wheat which also had Squarehead as one of its parents, while Juliana is a hybrid from Wilhelmina and another old English variety called Essex Smooth Chaff. As in the red wheats of the Squareheads Master type, so in the Wilhelmina group of white wheats, there is a number of other very similar varieties to the most commonly grown forms, and the two groups constitute between them a very high proportion of the wheat grown in this country. These white wheats are on the whole more suited to medium and heavy soils of good fertility, or to lighter soils in high condition in areas of relatively high rainfall. They should not be grown on the richest soils because of the danger of the crop becoming laid.

Disease resistance of wheat varieties is normally not one of the characters which determines the choice or preference by the grower for a particular sort for cultivation in this country, because those diseases which cannot be controlled by seed treatment or good rotations are usually not sufficiently severe to cause much worry. The commonly occurring Yellow Rust can, however, be responsible for material losses to susceptible varieties in seasons when the disease is especially favoured by the growing conditions. Some varieties are particularly susceptible to Yellow

Rust, others show some degree of resistance, while others have been bred specifically for the purpose of withstanding the disease and show considerable resistance. The first of these highly resistant varieties to be marketed in this country was Little Joss, a hybrid between Squareheads Master and a resistant Russian wheat named Ghirka. Little Joss has a red grain and chaff, a tall straw which makes it rather liable to lodge, and a capacity to ripen its grain evenly in spite of being laid. Because of its grain-ripening virtues and its resistance to Yellow Rust, Little Joss is grown to some extent on black fen soils, but it is essentially a variety suited to lighter soils of low to average condition. The newer variety Steadfast, derived from hybridising Little Joss and Victor, has a rust resistance equal to Little Joss, but is considerably more resistant to lodging because of its shorter straw; it possesses a white grain and is suitable for biscuit making as is Little Joss.

Each of the wheats so far described is suitable for autumn sowing and they include the varieties which are most commonly grown and those with special virtues for certain conditions and particular purposes. There are many other locally grown wheats of native origin in so far as they have been selected by breeders in this country, and there are also French wheats such as Bersée, and Vilmorin 27 which do well in this country. Swedish wheats, among which Iron, Steel, Scandia and Chevalier deserve mention, have had popularity in some parts of the country, but on the whole they have not proved so satisfactory for milling and blending with other wheats for bread flour as have the English and French wheats.

There is no difficulty in any part of the British Isles, where wheat can profitably be grown, in finding varieties of sufficient winter hardiness for autumn sowing, and the bulk of the crop is sown to the so-called winter varieties. This characteristic of British wheat growing is most important because it allows the cultivation of the slow-maturing high-yielding varieties; it enables the farmer to sow a proportion of the cereal crop in the autumn and thus spread labour; and it makes possible the use of the heavy soils, which are troublesome for cultivation and seed-bed preparation in the spring. On some of the intractable clays the old Rivet wheat is still grown, and a new selection of this variety termed Rampton Rivet, which is higher yielding and has a stronger straw than other strains, has done well on these soils and also on the chalky Down soils in the south of England. Some of the earlier maturing varieties of winter wheat such as Squareheads Master, Yeoman and Holdfast can be sown up to the middle of February, while others like Little Joss may even be sown rather later with reasonable chance of satisfactory crops. It is, however, more satisfactory to use the specially adapted spring varieties after the middle of February, although some of the winter varieties like Bersée may be sown in March.

Of the true spring wheats, there are some that are better suited to early sowing, and others which are satisfactory, but do not yield like the early-sown varieties, from April sowings. For many years the old French wheat Red Marvel was the commonly grown early spring wheat in this country, but in recent years improved varieties have been introduced. Outstanding among these spring wheats is the Swedish variety Atle, which like Red Marvel, and many others of the spring varieties in this group, has a white chaff and red grain. Atle, however, has a shorter and stronger straw and is higher yielding than Red Marvel, in addition to having the valuable attribute of being earlier ripening. Other suitable varieties for March sowings are the English hybrid variety Meteor (April Bearded × Yeoman), the Swedish varieties Blanka and Extra Kolben II, and the French Hybrid 29, which, however, is not a true spring form.

The choice of varieties for April sowings is limited, and for the most part there is little justification for sowing wheat as late as April, when the land would be better used for barley. The April wheats are all lax-eared, with red chaff and red grain, and their yielding capacity does not equal that of the varieties suitable for earlier sowing. For many years the standard variety for late spring sowing was April Bearded, but in recent years the stronger strawed Swedish varieties Fylgia and Diamant II have offered somewhat better alternatives.

The wheat varieties at the disposal of the farmer in this country offer a wide range of forms which differ from one another in many characters, some important and others of no practical significance. There are groups of varieties which resemble one another in all their essential visual and economically important characters, and there is every reason to reduce the number of such varieties. On the other hand, there is a nucleus of valuable varieties including Squareheads Master, Wilhelmina, Juliana, Yeoman, Holdfast, Redman, Warden, Little Joss, Steadfast and Atle which provide for the great part of the wheat acreage as far as growing conditions are concerned, and cater for all the demands of the market for home-grown wheat. The intrinsic worth of varieties depends on yield, grain quality, standing ability, disease resistance, capacity to respond to high farming conditions and environmental adaptation. These are the characters which should determine the farmer's choice, although a single character such as the greater resistance of most red-grain wheats to grain sprouting, during and after ripening in the field, may be the deciding factor in some circumstances.

The improvement of the wheat crop may be taken as typical of the plant breeding methods used also in barley and oats. Each of these cereals is self-pollinating, although occasional cross-pollination occurs. This means that single-plant selections may be used as the basis of all breeding work, and that varieties may be brought to a high degree of uniformity by selecting and building up from such single plants. But when the possibilities of improvement by such selection have been exhausted, future improvement is dependent on the creation of new types by hybridisation. In each of these three cereals the more recently introduced varieties are of hybrid origin derived from controlled and planned hybridisations followed by selection, a technique which is now the basis of breeding in these crops. Self-pollinating plants of this kind are, of course, easier to maintain true-to-type than is the case with cross-pollinating plants, and the multiplication of stocks and commercial seed production are not exposed to the same hazards of contamination from outside sources.

Chapter IX

BARLEY

It is still not known with any certainty whether barley or wheat is the more ancient of the cultivated cereals, but it is certain that both were cultivated in the most ancient civilisations of the Old World. Babylonian, Egyptian, Chinese and Indian centres of early civilisation had wheat and barley, and although barley was an important bread corn in certain areas, its use for making fermented drinks appears to be as ancient a practice, as far as it is possible to tell from recorded history. Some authorities believe that barley was used as a bread corn before wheat was, and that in some parts of the ancient civilised world barley was the more important crop for baking purposes. On the other hand, the oldest records of barley cultivation show that this cereal has been used for brewing purposes for many thousands of years.

With the discovery of the superior characters of wheat for bread-making, barley gradually gave way as a bread corn in all areas where wheat could be grown successfully, but it is interesting to see how far barley retained its position for bread-making in spite of the competition from wheat. Barley was highly prized as a bread corn by the Athenians, and it was considered next in value to wheat by the Romans, who preferred it as cattle food or for mixing with wheat to make household bread. Like wheat, barley appears to have spread with the growth of the temperate civilisations, its use being divided between bread-making, brewing and stock feeding, and the early history of its cultivation in this country seems to be similar to that of wheat.

Barley has a world-wide distribution as a farm crop of temperate regions very similar to wheat, but it is not grown as extensively nor does it have such importance in international trade as does wheat. The great value of barley as a crop for human and livestock consumption as a food, as distinct from a brewing material, is that it can be grown under certain conditions where wheat fails. Barley has a great range of successful cultivation because there is such a diversity of forms showing suitability for extreme climatic and soil conditions. There are both spring and winter types, some of the former having very short growing seasons, and some of the latter being very frost hardy. The quick-ripening types make it possible to cultivate barley at very high altitudes and as far north as the Arctic Circle. It is found in many mountainous regions of the world growing successfully at higher elevations than any other cereal, while its capacity to withstand heat and drought has enabled farmers in subtropical and semi-desert regions to cultivate barley successfully. But these are the more extreme types for which there is only a restricted and local demand, the extensive and most productive barley areas of the world are sown to other types which have suitability to better farming conditions and less extreme environments.

When barley was grown extensively as bread corn in this country it was found in practically every county including the far north of Scotland and the outlying islands. Its importance was greater in the areas where wheat cultivation was impossible, and we find that in the hilly districts of the north and the west, barley was the chief bread corn in the eighteenth and early nineteenth centuries. But in the lowland areas of the south and east parts of the country, large quantities of the barley crop were used for brewing and distilling, with substantial amounts also reserved for feeding all classes of stock. With the final change-over to wheaten bread in this country, barley has virtually lost all its value as a bread corn, except in wartime, and the crop is used almost entirely for brewing, distilling and feeding to livestock.

The important economic position of barley in the agriculture of this country is now primarily as a malting and brewing material, and there are definite and specific characters possessed by the grain of the barley plant which are responsible for this. Among the large number of different botanical types of barley is a great range of forms in which the grain actually adheres to the surrounding husk when the crop ripens. When threshed efficiently, the grains are completely enclosed by the husk, and no portion is exposed by breaking or tearing open of the husky envelope. This makes the various operations included in malting and brewing a much easier task. But in addition to this, the chemical composition of the food reserves of the grain, with its particular characteristics of a high proportion of easily fermentable starch and also its protein peculiarities, gives barley its special suitability for making a relatively cheap alcoholic beverage. There is no other crop which can be grown as easily or used so economically in this country for this purpose, particularly when the other ingredient of beer—hops—can also be grown successfully in certain areas.

Under the climatic and soil conditions of this country, barley has a wide range of cultivation, but good barley for brewing can only be grown in certain areas. There are very few soils on which barley cannot be grown, but undoubtedly the most suitable are the medium loams of free-working nature, or the lighter soils in good condition. Acid soils may cause crop failures, heavy clay soils usually cannot be worked to the desired kind of fine seed bed, and extremely rich soils lead to the crop becoming laid, a defect to which barley is particularly prone owing to the comparatively weak straw which characterises most varieties. All extreme conditions of both climate and soil lead to barley of poor brewing characters, although under such conditons good feeding barley may be obtained.

Owing to its particular characteristics and suitability, barley has become the most important spring-sown cereal in many parts of this country where good malting and brewing samples can be grown. But even in parts of the country where it is usually impossible to produce good malting samples, barley has a special place in crop rotations owing to its virtues as a late-sown spring crop. It used to be quite a commonplace practice to sow barley in May, while even June crops were sometimes taken, but in spite of the ability of barley to ripen with a shorter period from sowing to harvest than other cereals, it has now become customary to sow considerably earlier than this. The reasons for this are partly because higher yields can be obtained by earlier sowing with the better yielding varieties now available, and also the chances of obtaining a better malting sample are increased by this practice. Most growers of malting barley now follow the practice of sowing in spring as early as it is possible to obtain a good seed bed, and in good years this may be in February, although there is always some risk of damage by cold weather when a variety which is not frost hardy is sown as early as this. As a general rule the bulk of the barley crop is sown in March and the first part of April, but there is a great range in sowing dates from year to year, and it is fortunate that the best varieties are so accommodating. In some parts of southern England, spring varieties of malting barley are sown in the autumn, during October, or even in December, and although in mild winters this may result in earlier maturing crops of good quality, there is the constant risk of severe winter damage or outright killing of the young plants. To meet the requirements of farmers who wish to sow barley for malting purposes in the autumn, winter-hardy varieties are now being made available.

Something like two-thirds of the home-grown barley crop is used annually by the maltsters and the brewers who rely on the British farmer for the greater part of the raw material of this type, although some brewing barley is imported. Prices paid for malting barley are considerably higher than for feeding barley, and the total value of the malting barley crop runs into millions of pounds every year. The high cash value of barley as an industrial product—and in this connection it is important for the grower to market at the most appropriate time to obtain the highest price—is a very important consideration in the cereal-growing areas of this country where good malting samples can be grown, and all the important barley-growing districts are concerned primarily with supplying the needs of the maltster and brewer. Consequently although the maltster and brewer have adapted their processes to the kind of barley produced at home, the farmer has to study and understand the requirements of the trade if he is to take advantage of the good market at his doorstep, because the maltster and brewer have set standards which have decided what type of barley should be grown.

As far as the farmer is concerned, malting quality in barley differs from bread-making quality in wheat in two important respects. First, the varieties available and the growing conditions in certain parts of the country enable the grower to produce at home high yields of first-class material which is always marketable and often in great demand. Secondly, although the finest quality can only be grown by using certain varieties, this quality is very much at the mercy of the growing conditions and may vary very much from year to year even in the same district and on the same soil types. The great effects of growing conditions on malting quality have led to the impression by some growers that variety is of less importance, but this is not so. It is true that soil type, manuring, cultivation and management, climate and weather conditions determine very largely the malting quality of barley. That this is so may be seen from the fact that the best samples are usually grown in particular areas such as the south-eastern parts of England and certain coastal areas in Somerset, Devon and south-east Scotland. Also, the quality may be good, bad or indifferent as a whole practically throughout the barley-growing areas in any one season, while different parts of the country may be outstanding for good quality from season to season. But this does not alter the position that there are certain varieties which are not only the best farmers' varieties from the point of view of yield and straw strength, but they also produce consistently from season to season the best quality and are most in demand by the brewer.

There are certain tenets of good husbandry for growing prime samples of malting barley, and if these are not observed even the best of varieties under the most ideal conditions of soil and weather, will fail to give the desired results. The place in the rotation is important because a soil in too high a condition is apt to cause the crop to be laid and the grain becomes flinty, nitrogenous and discoloured. For this reason barley is commonly taken after a root crop or as a second cereal crop, but the soil type and fertility will influence the crop sequence. The best crops are usually obtained on a well-prepared and even seed bed, which gives ready and uniform germination and establishment of the plants. Heavy manuring, particularly with nitrogenous fertiliser, can lead to uneven ripening, rank growth, lodging, delayed ripening and a consequently poor grain sample. The crop should be allowed to stand until it is dead ripe, and it may be much improved by being allowed to mature in the field after ripening, although it may deteriorate if it is over-weathered by exposure to wet conditions. The grain must not be allowed to over-heat by stacking or bagging when damp because this may spoil the germination capacity on the malting floor, and threshing must be done so that the grain is not damaged by a close-set drum nor yet must pieces of the awns or beards be left on the grain. The grain sample should be even, well filled, of a good colour with a thin skin, free from mechanical and weather damage and white and mealy when examined inside. It is important that there should be no admixtures with other varieties of barley, or with another cereal, while dirt, weed seeds and other impurities are obviously undesirable.

The very many forms of cultivated barley in existence, and the difficulty of classifying these forms on any universally accepted system, has resulted in several suggested methods of classification and nomenclature. The genus *Hordeum*, to which the cultivated forms belong, includes several kinds of wild barley grasses, while there are also some primitive forms of grain-producing types not found in cultivation. It has been suggested that all the cultivated grain types should be grouped in one section of the genus, called *cerealia*, or alternatively that a single species—*sativum*—should be recognised. On the other hand, it is considered by some authorities that several cultivated species are distinguishable, although there is no general agreement as to how many.

The exact number of cultivated species of barley is an academic matter which does not materially affect the more generally accepted recognition and method of naming the important botanical and agricultural forms. This method is based on the obvious characters of the ear including the number of rows of grain; the colour of the chaff and the grain; the adherence or non-adherence of the chaff to the grain; the presence of beards (awns), or hoods, or the absence of any such appendage; and the denseness of the arrangement of the grains on the ear. All the varieties cultivated in Britain have white grain, white chaff, grain adhering to the chaff, and beards, but they differ in the number of rows of grain and the density of the ear. The forms with six rows of grain may be termed *polystichum* forms, and those with two rows of grain, *distichum* forms, there being both dense ear and lax ear forms in each group.

The requirements of the maltster and brewer, and the suitability of the growing conditions for the production of particular types of barley, have had far-reaching effects on the forms and varieties cultivated in this country. Although from time to time many botanical forms have been introduced, they have never had any great success and have soon been abandoned. All the varieties which have been extensively cultivated in Britain in historical times have shown a great similarity in their botanical type, and the crop is at the present time notable for the small number and uniform type of the predominant varieties.

The distinction between the six-row and two-row types of cultivated barley, and the recognition of dense-ear and lax-ear forms in each of these main types, is not only important botanically, but is necessary for a proper understanding of the utilisation of the cultivated varieties of this country. As long ago as the sixteenth century, there were references to the 'diverse kyndes' of lax two-row or 'English barley', cultivated most commonly in the southern counties, and to 'Byg or Beare', a lax six-row form confined mostly to northern England and Scotland. The 'English barley' is typical of the botanical type known as *nutans*, or nodding barley, and the 'Byg' (now known as 'Bere') is the characteristic 'four-row barley' of the botanical type *vulgare*. The corresponding dense-ear forms in the two groups are *erectum* (two-row) and *hexastichum* (six-row). Each of these botanical groups is still represented by agricultural varieties at present in cultivation, but their relative contribution to the total barley acreage of the country is one of the characteristic features of the crop to-day.

The various lax two-row varieties appear to have dominated much of the barley-growing area of England for hundreds of years, and many local varieties appeared. At some period, probably towards the end of the eighteenth century, a new type called Archer, or Archer's stiff straw, was selected, and it spread rapidly until it became the most widely grown form in the important barley-growing areas, but it obviously existed in a number of distinct types. There also occurred at this time, but more locally, dense-ear forms, the oldest and best known of which was Spratt, a variety famed for its strong straw and suitability for growing in the fens and on other rich soils. Both of these barleys have played an important part in the origin of present-day varieties, and the Archer barleys, which have been used most extensively in breeding, held the dominant position in this country until the early part of the nineteenth century and are still cultivated locally.

During the nineteenth century two chance selections gave rise to a lax-ear variety named Chevallier, and a dense-ear variety named Goldthorpe. These achieved considerable importance because of their improved malting quality, but Chevallier suffered from a weak straw, and Goldthorpe from a long 'neck' which led to considerable loss in the field from ear shedding. Both of these varieties are still grown to a limited extent, and their fame as malting barleys was so great that their names are sometimes used to-day to denote the two main types of malting barley—the lax ear and the dense ear. But at the beginning of the present century the dense-ear variety Plumage was introduced from the continent. This variety has been taken into cultivation in parts of northern England, where it has largely replaced Goldthorpe in certain areas because it has a shorter neck and is not so prone to ear shedding. Both Plumage, and a selection from it named Plumage 63, are earlier ripening varieties, and occupy a position of some importance in the cultivation of malting barley in certain districts of northern England.

During the period when Plumage was being taken into cultivation, two hybrid varieties were put on the market which have had a greater influence on barley growing and the malting and brewing trade than any other varieties. The first of these, a broad-ear variety named Plumage-Archer, is briefly an improved Plumage type; and the second, Spratt-Archer, is an improved Archer type. Both of these varieties, whose parentage is obvious from their names, have been so successful in combining the characters of a good farmer's barley with first-class malting and brewing characters that between them they contribute a high proportion of the total malting barley grown in England. They both have less tendency to lose their ears, better straw, higher yielding capacity and better grain characters than any other varieties which can be grown under similar conditions. Spratt-Archer is more widely grown and has a wider range of adaptability, but it is particularly suited to lighter soils and to the eastern parts of the country. Plumage-Archer is more successful on the heavier soils and is usually at its best in coastal areas and in the good barley districts of the more westerly parts.

Restriction of the number of varieties is very desirable for maltsters and brewers who learn the most economical methods of treating particular varieties, and it is obviously more convenient for them to have large quantities of a small number of varieties than small quantities of a larger number. But there is room for a wider range of barley varieties, particularly from the grower's point of view. Improved straw strength is urgently needed, and shorter straw for combine harvesting would be an advantage, while good early-maturing varieties would help to spread the labour of harvest and enable malting barley to be grown with greater chance of success in late districts. There are two Danish varieties named Kenia and Maja which possess these desirable field characters, but the grain is not the type most desired by maltsters and brewers. Efforts are now being made by plant breeders in this country to produce the desired types by hybridising the Danish barleys with the best of the English varieties, although the demand for earlier ripening has been met by the introduction in 1947 of the variety Earl which is a selection from Spratt-Archer.

Some farmers sow malting barley in the autumn, and it has always been customary to use the standard spring two-row varieties for this purpose. These varieties do not possess any real winter hardiness and are liable to considerable damage in severe winters or to considerable thinning out in less severe ones. To meet the needs of autumn sowing a new hybrid variety named Pioneer was made available to growers in 1943. This variety was bred by hybridising Spratt-Archer with a winter-hardy Austrian barley, and it possesses very similar characters to Spratt-Archer combined with a high degree of winter hardiness.

The cultivation of two-row malting barleys is the dominant and characteristic feature of barley growing in the British Isles, but as the whole of the crop is not used for this purpose, and for distilling, there is a considerable residue that does not reach the necessary standard and ultimately finds its way on to the market as a stock feed or else is used in this way on the producer's own farm. In 1945 a two-row barley for spring sowing, and more suitable for feeding than malting, was released to growers. This variety, which is named Camton, is a hybrid obtained by crossing Spratt and Archer-Goldthorpe. It has a broad and very dense ear and an exceptionally strong straw which makes it suitable for fertile soils or heavily manured soils which are capable of carrying heavy crops.

Six-row barleys are now only cultivated on a small proportion of the barley acreage and are used as a general rule for feeding stock. In upland areas of the north and west the old Bere is grown, and in Wales a similar type called Haidd Garw is still to be found. Although it is usually assumed that all six-row barleys are winter types, possessing winter hardiness and suitable for autumn sowing, this is not so, although the types grown in this country have usually been of this kind. The reason for the virtual eclipse of the six-row barleys in this country has been that the maltster relies on the British farmer to supply him with the two-row material and when six-row varieties are used in the trade they are imported. It is a very general practice in some breweries to mix imported six-row barleys with home-grown two-row varieties, and in some cases even foreign two-row material is also used. In an endeavour to meet the demands for six-row malting barley a new variety named Prefect was marketed in 1944, this variety being a winter-barley hybrid obtained by crossing Spratt-Archer with a six-row German winter barley.

Barley cultivation in the British Isles must continue to be primarily concerned with malting and brewing, although there is a large outlet for feeding barley. The breeding of high-yielding, high-protein barleys specifically for feeding is a problem engaging the attention of plant breeders, and such barleys may eventually assume an important place in the cereal growing of the country. Of the total amount of barley consumed in this country in normal times over 40 % is commonly imported, and although imported malting barleys may fulfil a purpose for the maltsters and brewers which home-grown barley cannot, there is no reason why considerably more feeding barley should not be grown. Barley meal is a particularly suitable food for pigs, while barley straw, although not of a high feeding value, is more palatable and nutritious than that of wheat.

The position of barley cultivation in the British Isles is assured on the malting requirements alone, but the suitability of the country for barley growing encourages the view that the crop could be grown more extensively for feeding purposes. Something like three-quarters of the crop is grown in England, and half of this acreage is in the extreme eastern counties, with additional substantial contributions from the south and south-eastern counties. In Scotland and Ireland there is a similar distribution in the eastern areas, and the whole localisation is due primarily to the trade in malting barley. Although there are these limitations to the distribution of the barley crop, when considered in terms of malting and brewing, there are extensive areas which could grow feeding barley. The country as a whole is well suited to the crop, as may be seen from the fact that there are few countries which produce higher yields, and there is no doubt that these yields could be exceeded if the aim were to produce feeding barleys.

Chapter X

OATS

Oats are not as ancient a cultivated crop as wheat or barley, and are not mentioned with the two latter cereals in the oldest records which have as yet been discovered. Before the Christian era, the only references to oats are as a weed in other crops, and it is probable that this cereal was eventually taken into cultivation because it became so prominent as a weed that it virtually usurped the place of wheat and barley in certain areas. The wild oat has been a notorious weed of farm land for many hundreds of years, and the difficulty of recognising it from cultivated wheat and barley when in the young stage was probably responsible for its reputation as a weed of more than ordinary powers. Wild and weed oats also have the characteristic of shedding their grain from the inflorescence when ripe, a habit which makes it particularly difficult to prevent the infestation increasing from year to year.

The way cultivated oats evolved and developed is suggested by what is known of the habits and characters of the many forms of wild and weed types. There are certain wild forms with small, pointed, heavy grains which possess strong awns or bristles arising from their backs by means of which they are helped in burying themselves in the uncultivated land on which they occur. These forms are not the weeds of farm land and are only known in the wild state. Other forms, with larger grains and more robust plants, have invaded cultivated land where they possess certain powers of 'crop mimicry', adapting themselves to the seasonal growth of the crop in which they occur, and eventually producing inflorescences with the crop, but shedding their grain before harvest.

It seems likely that the earliest oats to be used were these weed forms, which could be fed in the green state to livestock. Ultimately forms arose in which the grain was held tightly in the inflorescence on ripening, and these became the first true cultivated types. Occasionally plants are found in cultivated varieties, even at the present time, which have reverted to the

PLATES
10–18

HARVESTING WHEAT BY BINDER

PLATE 10

Deep, retentive soils of the type which will support good tree growth, are suitable for producing heavy yields of wheat. In the good wheat-growing districts of Britain, the crop may become too long in the straw, and in wet seasons there may be considerable difficulty in harvesting. By far the greatest part of the wheat crop, as of the total cereal acreage, is cut with a binder, although increasing use has been made in recent years of the combine harvester. The relatively damp atmosphere of this country causes the grain at harvest to have a high moisture content than in the drier climates of some of the big wheat-growing countries of the world, and when a combine harvester is used the grain must be dried artificially. Taking this country as a whole, the best wheat crops are obtained in the driest and sunniest seasons, and one of the limiting factors to more extensive wheat cultivation is high rainfall and cool conditions.

A FIELD OF RIPENING BARLEY

PLATE 11

Barley, like wheat, is a crop for low rainfall and plenty of sunshine, but it is grown most extensively on the lighter and shallower soils which are more suitable for the production of good malting samples and do not encourage too lush a growth which easily causes barley to lodge. Although barley used to be cultivated extensively as a bread corn in this country, approximately two-thirds of the crop is now used by the maltster and brewer, and the best grain samples are used in this way for making beer. Good malting quality is a varietal character, but the season and the soil are so important in their effects on quality that under unsuitable conditions the best varieties may produce grain quite unfit for malting. Barley which is unfit for malting is used for feeding live-stock, and in some parts of the British Isles the crop is grown entirely for consumption on the farm in this way, while smaller amounts are used for distilling.

A PURE STOCK OF OATS

PLATE 12

Oats are the most widely distributed and extensively cultivated cereal in the British Isles, and although small amounts are used for human consumption, the crop is grown essentially for livestock feed. Oat cultivation is, indeed, a reflection of the importance of the livestock industry in this country, and the crop is valued for the straw as well as the grain. Oats, being particularly suited to cool and moist climates, has special virtues as a cereal crop which are not shared by wheat or barley, while its ability to grow successfully on acid soils gives it special value. The use to which the crop is put, and its peculiar suitability to soil and climatic conditions unfitted for wheat or barley cultivation, give oats its special economic position in British agriculture. There is a large number of varieties suited to the very wide range of conditions under which oats are grown, while the feeding value of the grain, the proportion of grain to straw, and the suitability of the straw for feeding purposes are all varietal characters.

HARVESTING A SILAGE CROP

PLATE 13

Cereals, mixed with leguminous crops, are sometimes grown for cutting green and making into silage. One of the most widely grown mixtures for silage making in this country is oats, beans and vetches, and very heavy yields of green forage can be obtained under good conditions in this way. Silage making is a method of conserving green forage for winter use, and is extremely useful under a wide range of conditions in Britain. Not only may a balanced, succulent feeding stuff of high nutritive value be conserved in this way, but the method offers an alternative means of crop conservation to hay making, which is often difficult in this country. A wide range of crops can be used for silage, and the characters of the feeding stuff ultimately obtained depends on the crop or crops used, the stage of growth when cut, and the changes which take place in the silo.

CARTING SAVOYS IN WINTER

PLATE 14

The genus *Brassica* of the family Cruciferae offers a wider range of succulent food for human beings and animals than any other genus of plants cultivated in this country. The genus is outstanding for the number of different kinds of vegetative organs which become abnormally developed and remain edible for some time when kept fresh. The cabbage group of brassicas is remarkable for its leafy edible parts, and the cabbage itself is characterised by a swollen terminal bud growing on a short unbranched stem. The large number of cabbage varieties and types, and the adaptable nature of this cultivated plant, give this crop a very wide use. It may be grown for human or livestock consumption, and the many types make it possible by suitable management to have the crop maturing at practically any time of the year. Some, like the savoy, are useful for standing the winter, and may be fed on the farm or marketed as a vegetable.

PICKING AND BOXING BRUSSELS SPROUTS

PLATE 15

Brussels sprouts differ from the cabbage in developing small swollen axillary buds along the stem instead of the large single terminal bud. Sprouts are grown as a winter vegetable and are confined mostly to small holdings and market gardens in this country. The crop is essentially a cash crop, and requires intensive cultivation and much hand labour in picking and preparation for the market. There are considerable varietal differences in the form and shape of the sprout buds, and there are strong local preferences for particular types.

PACKING BROCCOLI FOR MARKET

PLATE 16

Cauliflowers and broccoli belong to the same species of the genus *Brassica* as the cabbage and brussels sprout, but the edible portion—the 'curd'—consists of an abnormally succulent inflorescence. The crop is grown entirely as a vegetable, broccoli differing from cauliflowers in being essentially a winter type and in having a slightly different form of curd. As in the case of the brussels sprout, cauliflowers and broccoli are cultivated mostly on small holdings and market gardens and require intensive management and handling.

WINTER FOLDING OF SHEEP ON LEAFY TYPE OF BRASSICA CROP

PLATE 17

Rapes and kales are leafy and succulent cruciferous plants whose chief value in this country is for stock feed, although some types of kale are grown as a vegetable principally in gardens and allotments. The commonest types cultivated in Britain are the swede-like rape, the thousand-head kale and the marrow-stem kale, all of which may be used for folding or cutting green. Rape is a smaller growing plant than the kales and produces a crown of large leaves: it is most commonly folded with sheep. Thousand-head kale develops an open branching crown, and marrow-stem kale has a long swollen stem which can be fed as well as the leafy crown which is a prominent feature of the crop. Rape is usually grown for consumption in the summer months and is a very quick growing plant, whereas the kales have a long growing season and are used for winter feeding.

PLATE 18

The swede, like the turnip, develops a food storage organ by the swelling of the tap root and the short stem below the seed leaves or cotyledons. Both plants belong to the genus *Brassica*, but are distinct species from one another and from the cabbage group. There are many types and varieties of swedes and turnips which offer a considerable range in growth rate, feeding value, keeping quality and disease resistance. As agricultural crops, both plants are grown almost entirely for stock feed, but small amounts are also grown as a vegetable chiefly on small holdings and in gardens. As a group, turnips are more quickly growing than swedes and are commonly folded: swedes, on the other hand, may be clamped for winter feeding. Both plants are seen at their best in moist and cool climates, where they constitute the characteristic fodder root under such conditions in this country.

wild habit of possessing grains which are easily shed, or which show a grain structure indicating this habit. There appears to be in oats a closer relationship between the weed forms and the cultivated types than in wheat and barley, although in the two latter cereals there are forms at present in existence which show wild or 'primitive' characters.

The general preference of oats for a milder and moister climate than wheat or barley, suggests that oats were not taken into extensive cultivation in the earliest civilisations of the hotter and drier parts of the world, although there are some forms which are better suited to heat and drought than others. There is evidence that the earliest intensive cultivation of the oat crop was in Western Asia and adjacent south-eastern parts of Europe, while there are also early accounts of oats as a farm crop in the cooler parts of China. The rapid spread of the types suited to the cooler and moister climates from the more southern area to the northern parts of Europe and Asia is a striking feature of the oat crop, and Bronze Age relics show the remains of oat grains, and are evidence of the use of this cereal in early settlements of Western Europe.

The distribution of the different botanical forms of oats in the Old World to-day shows their preference for particular growing conditions. The common cultivated oat which is familiar in this country is found throughout the cooler and moister parts of Europe, in Russia, south-west Asia and China. The bristle-point oat is centred more in north-west Europe, the Baltic coast, Portugal and locally in the western parts of the British Isles. The Byzantine oat is cultivated extensively in the countries bordering the Mediterranean, while north central China has a naked oat closely related to the common cultivated oat of Europe.

Oats are the most extensively and widely cultivated cereal in the British Isles, and wherever tillage is possible this crop can be grown. The acreage of oats formerly exceeded that of wheat and barley together, but the distribution is different from that of these two latter cereals. In Scotland and Wales, and the western parts of England, oats are grown to a much greater extent than they are in the rest of England, and although the total acreage in the

United Kingdom is greater than that of wheat or barley, oats are second to wheat in England. The reason for this distribution of the crop in the British Isles is largely climatic, but oats are also the most suitable cereal crop for the more infertile soils and where there is a tendency to soil acidity. The crop then can be grown economically where wheat and barley would fail, but excellent oat crops are found in areas where the other two cereals flourish, provided there is a good soil moisture supply. Under the latter conditions, the yields of grain are often very high on fertile soils, and the feeding quality of the grain is good: but under low rainfall conditions generally, the straw yield is lower than in wetter districts.

Oats are very accommodating with regard to climate and soil and are grown economically further north in some parts of the world than any other cereal. The large number of distinct varieties showing preference for particular climates, soils and levels of fertility makes this wide distribution of oats possible. The high water requirement of the crop results in it being most productive on deep retentive loams and clay loams, or where there is a high organic matter content in the soil, and peaty and moorland soils can bear excellent crops provided any mineral deficiencies are made good by application of fertilisers. These particular characteristics of oats make it a very suitable crop to take after ploughing up grassland of various kinds, and it is for this reason that oats are commonly the first crop after grass in all parts of the country. Two oat crops are often taken in succession in this way, and the crop is sometimes used entirely as a means of putting the land through a short rotation in between ploughing out worn-out pastures and reseeding. It is a common practice to sow oats after root crops, or to take it as the second white straw crop after wheat. The fact that oats are most commonly spring sown affects its place in the rotation, and where autumn sowing is practised its position may be changed.

Oats are used primarily as a stock food and are mostly home-consumed, there being little international trade. This country normally imports a small amount of oats, but of the three important cereals consumed in Britain, the nearest approach to self-sufficiency is reached with oats. A large proportion of the home-produced crop never leaves the farm where it is grown, but is fed directly to the farmer's own beasts. Some of the best grain samples are sold in the open market as stock feed, and a small proportion is used for the preparation of oatmeal for human consumption. As in the case of barley, the use of oats as a human food has shrunk to inconsiderable proportions with the increased consumption of wheaten bread, but whereas barley is normally never used in this country for baking purposes, oats are still employed for making oatcakes, and other cooked cereal foods.

It is then as a stock food that oats are of such importance in the economy of British agriculture, and the crop has especial value by virtue of several methods of utilisation. The grain, either whole or crushed, may be used liberally for feeding horses, sheep and all classes of cattle. It is a safe, nutritious, and, apart from the husk, an easily digestible concentrated food which is very palatable to animals, and seems to have beneficial effects which are not conferred by wheat or barley. The composition of the grain without the husk gives a well-balanced food with a higher proportion of oil than in any other home-grown cereal, and although the composition varies to some extent between varieties and under different growing conditions, this character is not taken into consideration in the choice of varieties. The reason for this is that the most important character affecting the feeding value of the grain is the proportion of husk which varies considerably not only between varieties, but with the growing season, while the composition of the husk varies with the ripeness of the grain at the time of harvest. All these grain characters affect the milling quality of the grain for oatmeal manufacture; the proportion of husk, the water percentage, the colour, flavour and keeping powers all being taken into consideration.

Apart from the great value of the grain for feeding, the oat crop has a special place in British agriculture because of the extensive use of the straw as animal fodder. Oat straw is valued above all other cereal straws for this purpose, because it possesses special virtues as far as its composition is concerned that are not possessed by the other cereals, partly because of the management of the crop. Oats are commonly cut before the crop is fully ripe and while there is still considerable nutriment in the straw, which becomes a poorer and less digestible fodder as the crop ripens. Although the feeding value of the straw is largely a matter of the stage of ripening, it is also affected by the growing season and the variety, and certain varieties are valued principally because of their high yields of straw which can be harvested in a nutritious condition.

Oats can also be used for making into hay, but in this country they are very seldom used for this purpose. The principal utilisation of oats as a green forage crop is for silage, for which method of handling they are usually mixed with a leguminous plant such as peas, beans or vetches, or a mixture of two of these legumes may be used. Such silage mixtures, when properly balanced with regard to the proportion of oats and the leguminous plants, can give extremely valuable stock feed if cut at the right time.

The genus *Avena* to which all wild and cultivated oats, as well as certain grasses, belong, includes three important species as far as the cultivated oats of the world are concerned. The most important and extensively grown varieties in cultivation to-day, not only in Britain but also in the cool temperate regions as a whole, belong to the species *sativa*, which contributes all the high-yielding, grain-producing types. *Sativa* oats are grown under the higher fertility conditions and in lowland areas of this country, and as a whole may be said to characterise the important oat-growing areas. In the early days of oat cultivation in this country it seems probable that the forms grown were largely of the species *strigosa*, which is better adapted to poor soils, and wetter climatic conditions. At the present time there are still some *strigosa* ('bristle point') forms grown, particularly in the upland and hilly districts of Wales, and also in Scotland and the Shetlands. The third important oat species—*sterilis*—is not cultivated in Britain, although there are some recently bred hybrid varieties now being grown which are derived from hybridising *sativa* and *sterilis*.

At one time, naked oats appear to have been widely cultivated in Britain, and a relic of these times is still found in Ireland under the name of Piley. This is a *strigosa* oat, and is a very localised survivor of the type of oats cultivated by certain of the Celtic peoples of these islands. The *strigosa* oats, with some of the naked types, survived as an important crop in late and wet districts for many hundreds of years because of their special suitability to these conditions where they are still the only type which can be cultivated economically. Naked oats are now virtually obsolete in Britain, for all practical purposes, all varieties having grain invested in the husk, which, however, does not fuse with the grains as it does in barley.

The very wide range of soil and climatic conditions under which oats are grown in Britain, and the different emphasis which may be placed on the relative value of the grain and straw, or whether the crop is being grown for green forage, naturally results in there being a large number of varieties in cultivation. It is therefore convenient to group the varieties according to important botanical and agricultural characters, both for clarity of description and for consideration of the economic value which may be ascribed to each variety. The older varieties cultivated in this country are forage oats which produce relatively high proportions of straw to grain; they are suited to lower fertility soils and more exposed situations with less favourable climates for grain production. This type of variety is often described as a 'straw producer', in distinction to the more recent 'grain producers' which are capable of giving higher grain yields under more fertile and more favourable conditions. This distinction is very important because it is the first consideration in deciding what type of oat to grow under any particular set of circumstances.

The second point to consider is the season of growth and suitability for spring or autumn sowing. By far the greater number of varieties are suitable for spring sowing and there is a larger choice in this group than in the winter forms. The spring varieties vary considerably in their time of ripening, there being early and late forms, and even comparatively small differences may be of great importance in late districts where a few days may make all the difference between a successful or difficult harvest. Winter hardiness does not reach such a high degree in oats as it does in wheat and barley, and even the most winter-hardy varieties are liable to be damaged in severe winters. The lack of winter-hardy material in oats is very marked, and is the reason why there are so few varieties characterised by winter hardiness at present available to growers. This distinction between winter and spring varieties is very marked and most important for the grower, who should realise that some varieties which are described as 'winter' possess very little in the way of winter hardiness.

Grain characters in oats are important economically only in so far as they affect the feeding quality or suitability for making oatmeal. Differences in the chemical composition of the 'kernel' have already been referred to, as also have differences in the percentage of husk. It is the latter which is the principal criterion in judging a good or bad quality oat, and it should be borne in mind that high quality means low husk percentage. Such superficial characters as the colour of the husk are un-important, as also are the size and shape of the grain unless these latter are such as to have a material effect on the husk percentage. White-husked oats are the most popular in the open market, but yellow, grey or black may be as good as, or even better, in feeding value. A well-filled grain is desirable under all circumstances, but some very plump grain varieties have a high husk percentage, while some of the boldest and largest grain varieties are the huskiest of all. It is well, therefore, not to assess the intrinsic value of oats on the colour, shape and size of the grain.

The other characters which distinguish oat varieties are the type of inflorescence or panicle and the length and standing capacity of the straw. The shape, size, conformation and attitude of the panicles show a great range among the many varieties and it is usual to characterise them particularly as 'equilateral' or 'unilateral' (one-sided), according to whether the branches arrange themselves equally around the main axis, or whether all lie on one side. The branches may be drooping, horizontal, inclined or practically erect: they may be long or short: and they vary in number. Some of these panicle characters are associated with particular kinds of straw such as long open panicles with tall straw and erect branches with stiff straw, but there is such a wide range of straw length, strength, coarseness and fineness that it is impossible to characterise them in anything other than the most general terms.

The development of improved varieties of oats in this country is generally considered to have started with the introduction of a variety called Poland in the early part of the eighteenth century. Various selections were made from this variety, one of the best known being Tam Finlay, and during the remainder of the eighteenth century several new varieties such as White and Black Tartary, and Potato, appeared. The last-named variety is still in cultivation and is a good example of the great and unforeseen success which can attend a chance selection found by the discriminating and observant agriculturist. These improvements were followed in the nineteenth century by still more selections, some as the results of chance, and others as the result of careful and organised work. Among the more important of these were Sandy, Hopetoun and the Fellow oats, most of which have now gone out of cultivation. In 1892 the modern era of oat breeding saw its beginning with the appearance of the hybrid variety Abundance, and from this date until to-day there has been a steady and prolific supply of new and improved varieties which have done so much for oat cultivation in this country.

Most of the improvement in modern times has been with the grain-producing types for growing on soils of good fertility. Where straw-producing varieties are grown, and on the poorer soils, particularly in upland districts, the older varieties such as Tam Finlay, Sandy, the Welsh bristle-point varieties (Ceirch Llwyd and Ceirch Llwyd Cwta) and Potato, which is not an extreme straw producer, are still grown. These varieties, with Black Tartarian, are suitable for the high rainfall and late districts in the west where they are used for producing high quality leafy forage which may be cut green, although some have good quality grain as well.

But the most widely grown and profitable oat varieties in the country are the grain producers with white grain and equilateral panicles, suitable for good lowland conditions. Many of these have very strong straw and can be grown on highly fertile soils, but the grain quality is often not the best from the point of view of the husk percentage. The majority of these varieties are suitable only for spring sowing and do not possess any winter hardiness. Two of the most popular varieties in this group are Abundance (White Swedish × White August) and the Swedish selection from the old Milton oat named Victory, both of which are good standard varieties for average to good soil conditions. There are more recently produced hybrids available which can be grown under most conditions which are suitable to Victory, but which are superior to this variety in certain respects. Eagle (Von Lochows Yellow × Victory) is a higher yielder with a rather smaller grain of better quality, but ripening later than Victory, and Star (Victory × Crown), which shows a slight all-round superiority to Victory, are valuable introductions from Sweden. In the western parts of the country S 84 (Victory × Red Algerian) and in Scotland Early Miller (Potato × Record) do well partly on account of their earliness. These two last-named varieties, with Eagle, Elder (Potato × Beseler's Prolific), Ayr Bounty (Potato × Yielder) and Resistance (Grey Winter × Argentine), are suitable for the most fertile soils because of their very strong straw, although Early Miller is not as good in this respect as the others.

In some parts of the country unilateral panicle types are popular, partly because of their conspicuous inflorescences. Some of these varieties are poor in feeding value, both with regard to grain and straw, but they have definite good characters in their resistance to lodging, high yields under fertile conditions, and earliness. Typical examples are Yielder (Waverley × Tartar King), Marvellous [(*Avena fatua* × Grey Winter) × (*Avena fatua* × Goldfinder)] and Onward (Marvellous × Superb). These varieties, with those previously mentioned, offer a considerable range of types to the grower, but they by no means exhaust the useful varieties available. Record, the two yellow grain varieties Goldfinder and Golden Rain II, the black grain Black Supreme, Engelbrackt, Black Bell III and Radnorshire Sprig, all have their virtues and adherents in localised areas in various parts of the country.

Winter oats are not widely grown in this country and the

choice of varieties is very much more restricted than is the case with spring oats. The advantages of autumn sowing are that the crop ripens earlier and is much less liable to damage by the Frit fly, which can cause serious loss in some districts especially when spring oats are sown late. Early ripening may be a disadvantage, when there is only a small amount of the crop in a district because of damage by birds, and greater resistance to Frit fly in spring varieties is being sought by breeding specifically for this purpose. The spring variety Eagle shows more resistance to this insect than any other variety at present in cultivation.

The oldest winter varieties cultivated in this country are Grey Winter and Black Winter, both of which appear to have been introduced from Europe many years ago. Neither of these varieties possesses sufficient frost hardiness to withstand the severe winters which are experienced in this country from time to time, but Grey Winter is the hardier and has a greater range of successful cultivation. Both varieties have good quality grain, the one of a pale grey and the other black, and although their yields are fairly good, the straw is weak and very liable to lodge. Attempts to breed improved winter-hardy varieties have all centred on hybridising Grey Winter, which is the hardiest oat available, with spring varieties in an endeavour to produce a higher yielding, stiffer straw type possessing a white grain and with the quality and frost resistance of Grey Winter.

The first of these winter hybrids to be put on the market was Bountiful which is a complex hybrid with Grey Winter and Black Winter blood. Although having a strong straw, and good yielding capacity, the grain is black and the frost resistance not equal to Grey Winter. Like all the winter varieties, it has an equilateral panicle. The variety Unique (Grey Winter × Grey Winter) was the next winter hybrid to appear, and it possesses all the virtues and faults of Grey Winter but has a white grain. More recently S 147 (Grey Winter × Marvellous), S 81 (Kyko × Grey Winter) and S 172 [(Kyko × Grey Winter) × (Bountiful × Grey Winter)] have shown a combination of increased straw strength, white grain and frost resistance rather below that of Grey Winter in some parts of the country. S 172 possesses a very short, stiff straw and is suitable to very fertile soils, while

S 147, which is susceptible to eelworm, is of very good grain quality. The most promising variety for low rainfall conditions combining the frost resistance of Grey Winter with a white grain of good quality, high yield and short, stiff straw is the new hybrid Picton (Grey Winter × Argentine).

The varieties Resistance and Marvellous, which were mentioned under the spring oats, are sometimes regarded as winter forms. While they are suitable to winter sowing in sheltered districts, they do not possess sufficient frost resistance to be regarded as safe for autumn sowing. They do well, however, from early spring sowings, when Resistance particularly is capable of very high yields on fertile soils.

Although oat cultivation is concerned essentially with the production of livestock food, the crop even from this point of view is dual purpose in that the grain and the straw are the primary products. Owing to the very considerable variation between varieties in their suitability for use as producers of grain or of straw and the marked varietal adaptation to soil types and climatic conditions, the varietal question in oats is in many ways more complex than in wheat or barley, even when the crop is to be consumed on the farm. This complexity is more obvious still when other characters are taken into consideration, and when in addition there is the problem of oat grain for sale off the farm to be considered. Fortunately, the varieties that give proportionately high yields of straw to grain possess good feeding quality straw, but although the grain is also of good quality, it is small, often of a type not popular on the market, and of course not produced in great quantity. There is no variety that combines the best features of a straw producer and a grain producer, and this fact, combined with the high tillering capacity, and suitability for poor soils and high rainfall of the straw producers, in contrast to the low tillering and high fertility requirements of many of the grain producers, has much to do with varietal distribution.

The whole tendency of oat cultivation in this country is towards the higher yielding and more profitable grain-producing types, which have straw of reasonably good feeding quality, particularly when grown in the wetter and cooler districts. The old straw-producing varieties must continue to be grown in the late and wet districts where the soil is poor, and some of the finer straw-grain producers are still the most suitable for poor upland conditions. But the most widely grown standard varieties which contribute the great bulk of the oat crop are the new high-yielding types with the attractive grain. It is a peculiar position that often the most profitable oats for sale are those of indifferent feeding quality, merely because oats tend to be bought on appearance, and the finest looking grain may have the highest husk percentage. The problem for the plant breeder is to provide varieties with sufficiently strong straw to stand up under the highest yields, while at the same time retaining feeding value in both grain and straw and not sacrificing unduly the demand for a large and well filled grain.

Chapter XI

RYE AND MIXED CEREAL CROPS

Rye, like oats, appears to have been first taken into cultivation some 2000 years ago, and the indications are that this occurred in Asia Minor. All the known botanical forms of rye are found growing in a comparatively restricted area of Asia Minor, but as in the case of the other cereals, rye has now spread far beyond this centre, where occurs a number of types with brittle ears, grains held firmly by the chaff and perennial habit, all characters which are regarded as primitive. Compared with the other temperate cereals, rye is much more limited in the number of botanical forms, and although several species are recognised, all the important cultivated forms are included in one comparatively uniform species—*cereale*—of the genus *Secale*, which comprises both wild and cultivated forms.

As an agricultural plant, rye has become essentially a crop of the north temperate regions. It has a wide range of cultivation, both with regard to climate and soil, and can be grown successfully under conditions which are quite unsuitable to other cereals. It is very hardy in that there are drought resistant and frost resistant forms which can be cultivated economically where no other cereal would be suitable, and although it does best on good soils, it can withstand acidity and soil impoverishment, particularly on light, dry soils, better than wheat, barley or oats.

These characteristics, combined with the fact that rye can be used for making bread which is nutritious and much liked by some peoples, have made rye an important cereal for human consumption in certain parts of the world. Rye is the dominant food grain only in a few countries of central and eastern Europe, but it shares first place with wheat in other European countries, and is cultivated most extensively in the U.S.S.R., Germany, Poland, Scandinavia and most central European states. It is only in those countries which have retained rye as a bread corn that it has continued to occupy an important place in agriculture, although there are restricted and localised areas of rye cultivation in some parts of the world where it is grown for making biscuits.

In the British Isles, rye has ceased to be a cereal of any great importance since the change to wheaten bread in the eighteenth century, and by the beginning of the nineteenth century the crop was valued principally as a green forage for stock feeding. In 1939 there was only 18,000 acres of rye in the United Kingdom, and the great proportion of this was grown in the eastern and southern counties of England. In addition to the value of rye as a cereal grain and a green forage crop, its long, strong straw has special virtues for thatching, making containers for bottles and other articles, and for packing, but the crop could hardly survive as a useful agricultural plant on the uses to which the straw may be put.

Compared with other cereals, rye has few named agricultural varieties, and these varieties are usually not as uniform in their characters as is the case with other cereals because cross-pollination is the rule in rye but the exception in wheat, barley and oats. Although a number of new named varieties have been introduced on the continent, there has been very little breeding of rye in this country because of the unimportant character of the crop. The choice for the British grower usually lies between the winter varieties known as Common, Star, Giant and Russian, or a spring variety such as Spring Common or Midsummer. Recently the Swedish variety King II has been made available in this country. This variety has a shorter straw, stands better and has a broader ear than the commonly grown varieties. The winter varieties are all very hardy, tillering profusely in the autumn and spring, and give higher yields of larger and better quality grain than the spring varieties. Autumn sown rye may take the place of any cereal crop in the rotation and when taken for grain usually follows clover or roots. As a catch crop, autumn sown rye is folded off in the spring and followed by roots. Spring varieties do not tiller and produce the large amount of early green forage which is the great virtue of autumn sown varieties, but a late sown crop of Midsummer rye may be folded or cut green in the autumn of the seeding year and again in the following spring.

Rye sown as a pure crop may therefore occupy one of several places in the rotation and can be used in more than one way. One of the ways of managing the crop is to fold it in the young stage and then take a grain crop. For this purpose the crop may be sown in the summer, fed off in the autumn and spring, or alternatively sown in the autumn and grazed only in the spring. The use of rye for supplying green herbage is made possible by the very vigorous leafy growth of the plant in its early stages of development, and mixtures of rye with rape or turnips are sometimes sown for grazing purposes. Such mixtures can supply very valuable stock feed in the late autumn and early spring if the weather conditions are not too severe.

In normal times approximately half the rye acreage is used for cutting green and silage making. As a constituent of silage mixtures for autumn sowing on poor, light land, rye may be used with vetches, but care must be taken not to leave the cutting too late because the rye becomes fibrous comparatively early in the season, and usually before the vetches are at their best.

During times of war, rye increases in importance in this country and considerably larger acreages are grown because land is brought into cultivation which will not profitably bear any other cereal crop. Most of this increased production is for grain, and the greater proportion is used for milling flour to mix with wheaten flour for bread-making. The rest of this wartime grain production of rye is used for stock feed, which is the most important form of utilisation in normal times, but even so, is a very small contribution to the total stock feed of the country. Even the relatively large increase in the cultivation of rye in this country during wartime still leaves it as insignificant compared with the peacetime production of any other cereal, and it is inconceivable that the crop will ever again play an important part in British agriculture. Its permanent position will depend largely on how far it will be economic to cultivate soils to which rye is peculiarly adapted, because it cannot compete with the other cereals in their particular spheres. There will always be some use for rye as a forage crop, and also for the manufacture of crisp breads and dry biscuits when imported supplies are not available, but these uses combined with grain feed for stock do not amount to a really important productive asset to the agriculture of the country.

In parts of the British Isles, particularly in the south-west parts of England and in isolated areas in Wales, mixed cereal crops are grown on a restricted scale. The commonest type of mixed cereals consists of barley and oats and is usually referred to as 'dredge corn'. This mixed crop is grown entirely as a grain feed for stock, and it commonly takes the place of oats in the rotation. The advantages which are claimed for this practice are that the yield of grain is usually higher than from either cereal by itself and a better yield of more nutritious straw, which stands rough wet weather more satisfactorily, is often obtained. Wheat is sometimes added to the oats and barley, and the combined grain product, which can be varied in composition according to the relative amounts of each sown, gives a stock feed which can be fed to all classes of animals. This method of growing cereals is most common in Cornwall, and is usually explained on the score that it minimises the risk of crop failures due to the uncertainties of climate and lack of specific knowledge on soil requirements.

The practice of sowing mixed corn crops does not materially affect the economic position or agricultural value of the individual cereals as cultivated plants in the agriculture of the country. The mixed corn crop is primarily a device to minimise risks and obtain a convenient blending of cereal grain for stock feeding. Most farmers prefer to keep the cereal crops separate, and this is certainly the prevailing method of cultivation in all the arable districts and the areas of intensive mixed farming and other areas where cereals are grown most successfully and under the most suitable conditions. The obvious strongholds of mixed corn are where cereals are regarded essentially as stock feed for home consumption, and in the nature of the case, there can be little to recommend the practice to the grower, more happily placed, who can concentrate on a higher level of production of the individual cereal crops.

Chapter XII

LEGUMINOUS PLANTS: SEED AND FORAGE TYPES

There is no single botanical family of plants which is so important to the agriculture of the world as the Gramineae, whether considered as the source of food for direct human consumption or as the principal means by which livestock are maintained on the farm. The cereals and grasses are pre-eminent chiefly as a source of carbohydrate for man and animals, as well as supplying fodder, and a comparatively rich protein feed when fed to stock in the young herbaceous conditions. Agriculture without these plants is difficult to visualise; but with intensive food production it is necessary to look to another family of plants, which occupies a place second only to the Gramineae, because it supplies a type of concentrated food and herbage which is not given by the grasses and cereals.

This family is the Leguminosae, sometimes called the 'Pea Family' because of the general familiarity of the various forms of peas which are characteristic members. The family is very widely distributed throughout the world, is one of the most extensive of all families of plants, and is extremely variable in habit of growth, there being trees, shrubs, climbers and small herbaceous types. The economic and agricultural value of these plants depends on their capacity to produce proportionately a greater amount of protein than any other vegetable food, and various parts of the plant are utilised as cheap forms of protein. The most widely grown forms are herbaceous annuals or perennials, but in tropical and subtropical regions shrubs and trees are used.

As a food for human beings the seeds and pods are exclusively used. Seeds of leguminous plants, fresh or dried, have been used from very early times as the cheapest source of protein available, and they are a virtual necessity when animal products, particularly meat, are scarce or unobtainable. Dried bean or pea seeds were a staple part of the diet of the mass of the people in this country and continental Europe before meat and dairy products

96

were available at reasonable prices, and also before the potato became popular. The succulent green pods with the immature seeds, or the immature seeds by themselves, are a more recent development resulting from the increased recognition of the value of fresh vegetable products.

Leguminous food for livestock is confined to three main types. The ripened and dried seed may be fed after crushing or breaking; the partially mature plant with pods and seeds may be fed green or ensiled; the herbaceous part may be made into hay, cut green or grazed. The most valuable forage forms are rapidly growing annual or perennial plants which produce edible seeds or abundant and succulent vegetative growth rich in protein and certain minerals. The different kinds of growth habit and life-span in the various genera and species make it possible to choose an appropriate leguminous plant for several types of management and utilisation. Certain annual forms like peas and beans can be taken as a major crop in normal rotations, sown by themselves or mixed with another crop. Quick-growing annuals occupy the place of a catch crop and can be used for folding or cutting green as with black medick and crimson clover. The more robust forms of perennial forage types, like lucerne, may be sown alone or mixed with one or more grasses and kept down for a number of years to supply green forage for grazing, cutting green or hay. The smaller herbaceous forms such as the clovers are used primarily for mixing with grass to produce all kinds of swards from one year to permanent, the herbage being utilised for grazing or cutting.

Apart from the special value of the products of leguminous plants, and the important position which they occupy in agriculture by reason of the proteinaceous food they supply, there is a further matter of practical importance which enhances the value of these crops, and no economic consideration of this family is complete without some reference to their important nitrogen-fixing powers through the activity of the nodule-forming organisms which live in association by entering the roots of these plants. This organism, known as *Rhizobium leguminosarum*, occurs in the soil and is capable of living in this free state for some time, but it is only capable of maintaining its most active existence indefinitely if it can penetrate the roots of leguminous plants where it lives in harmony with its host. The presence of the organism stimulates the production on the root of small gall-like bodies, called nodules, which are the centres of the organism's activity. The rapid multiplication of the organisms results in an increased rate of fixation of the free nitrogen in the soil, the organisms performing this function only when they have a readily available supply of carbohydrates, which in this case they obtain from the host plant. In this way the soil is enriched in nitrogen by the decomposition of the root residues in the soil with the disintegration of the nodules themselves, but it is not known to what extent there is actual excretion of the nitrogenous material into the soil during active growth of the crop and development of the nodule.

This method of enriching the soil in nitrogen has been known at least since Roman times, and the use of leguminous plants for this purpose has been a recognised practice in this country since rotational farming became established. There are numerous strains of the organism which show preference for particular species of leguminous plants, and although the same strain may operate successfully with more than one host species, it is possible when a new leguminous species is being introduced on the farm that the correct strain may not be present in sufficient amounts for the successful growth of the crop. This may be the case with lucerne, for example, and it has become the practice to inoculate the seed with the appropriate strain of the organism before sowing the crop.

The peculiar relationship of leguminous plants with the nitrogen-fixing organism not only makes it possible to use these plants to enrich the soil in nitrogen, but it enables the grower to cultivate certain leguminous crops on particular classes of infertile soil. It is only when living in the tissues of the host plant that the nitrogen-fixing organism finds the conditions suitable for its activities, and it is only when the leguminous plant is invaded by the organism that sufficient nitrogen is made available for the host plant to grow luxuriantly, particularly on infertile soils where nitrification is liable to be inadequate through the activities of free-living bacteria.

This peculiar form of nutrition shown by leguminous plants explains the capacity of some species to grow naturally on infertile soils, and also the special value of the cultivated plants in this family for building up soil fertility. Most of the crop plants of the Leguminosae cultivated in temperate regions prefer calcareous and alkaline soils, but a few, such as the lupin, grow more successfully under non-calcareous conditions. There is, however, considerable variation in the soil types on which leguminous crops can be grown, some showing preference for clays, others for loams, and a few being particularly suitable to the lighter classes. But in all cases the plants can be stimulated in their growth by adding phosphates to the soil, because this has a stimulative action on the nitrogen-fixing organism, increases nodule formation, and thereby results in a more luxuriant growth of the crop.

It is commonly stated that leguminous crops are good drought resisters because they produce deeply penetrating tap roots. This is true of certain herbage and forage types like lucerne, kidney vetch and common vetch, but it by no means applies to all. Free drainage is certainly an essential requirement for most leguminous crops, but the water requirements of individual species, and their capacity to make sufficient growth for economic cultivation, varies considerably. It must also be apparent that conditions that are suitable for the cultivation of a particular species for herbage or green forage, are not the same as for its cultivation as a seed crop. Thus peas may be grown for forage over a considerably wider area than peas which are being cultivated for the mature seed, and conditions which are admirably suited to grazing legumes are not necessarily suited to the hay types.

Field beans and peas grown to maturity on the farm are often classed agriculturally as 'grain crops', but this is rather misleading and confusing, because there is no botanical association or relationship between the true grain crops, or cereals, and beans and peas, while the agricultural management, place in rotation and utilisation are quite distinct. It is not uncommon, also, to refer to the clovers used in herbage mixtures for grass swards as 'grasses', but again this mental association of a leguminous plant with a graminaceous plant should be avoided because of the entirely distinct characters of the two families. In this country, and most temperate regions, all the leguminous agricultural plants belong to one section of the Leguminosae called the Papilionaceae, because they possess the characteristic 'butterfly' shaped flower so well known and easily recognisable to gardens, farmer and field botanist. It is usual, therefore, in this country to think of all leguminous plants as having the characters of the papilionaceous forms which are so familiar, and to refer to them in the more general term as leguminous plants.

The common leguminous seed crops cultivated in Britain are the field bean, and the field and green pea. The common field bean, which is merely the agricultural form of the broad bean of the garden, but considerably more ancient as a cultivated plant, is an annual plant belonging to the genus *Vicia*, which includes also the vetches. *Vicia faba*, the common bean, probably belonged to all the ancient agricultural civilisations of the old world, was certainly cultivated before the Christian era, and was well established in Ancient Egypt. The Greeks, Romans, Chinese, Japanese and inhabitants of Northern India used the bean as a vegetable or field crop, and it has since spread throughout the temperate regions of the world. It is usually assumed that the Romans introduced the plant to this country and that one of its botanical names is derived from Fabius, the Roman patrician.

The agricultural value of the field bean is due to the concentrated proteinaceous character of its seed which contains something like 25 % of reserve protein. With the field pea it is the only protein concentrate of this type which can be grown as a stock feed in this country. During the present century the cultivation of this crop has decreased considerably on British farms, in fact the acreage devoted to beans has suffered considerably from the introduction and extended cultivation of other crops, notably clover, roots and potatoes. In the middle of the last century over half a million of acres of beans were being grown in the United Kingdom, but by 1939 there were only 135,000 acres, most of which were concentrated in the eastern parts of England, and confined largely to the heavier soils. The greater proportion of the crop is grown to maturity for the seed when the rest of the plant is of little value for feeding, but a small proportion of the crop, principally in the north, is cut green or in the partially ripe condition when the straw is then valuable. Beans are also grown with cereals or another leguminous crop such as vetches for silage, and the horticultural types which have larger seeds of a white or green colour, instead of the chocolate brown of the field bean, are used for picking green for human consumption.

It is on the heavier classes of soil that field beans are seen to the best advantage, and their cultivation is largely confined to such soils at the present time. They are generally taken between two cereal crops, or they may occupy the place of clover in the rotation. South of the midland counties of England, the crop is commonly sown in the autumn because, although not sufficiently winter-hardy to withstand severely cold conditions, winter beans are more successful in drier districts. Also, winter beans do not suffer so badly from 'black fly', and autumn sowing is more easily achieved on heavy soils, than is spring sowing. In the northern parts of England, and in Scotland, spring sowing is the usual practice, but in general, beans are a much more successful crop from autumn sowings and they prefer milder climates, sunshine and a not too high rainfall provided there is ample soil moisture.

This matter of the adaptation of the crop is very important, because at the best of times, and under the most suitable conditions, beans are an uncertain crop, the yield fluctuating very considerably from year to year. 'Good bean years' are not common, and one of the most unsatisfactory features of this crop is its uncertainty, a condition which has had much to do with the contraction of the acreage and a greater reliance on the part of the farmer on imported concentrates. Concern over the position of the bean crop has stimulated research on the cultivation, fertiliser treatment and possible methods of improvement of available stocks in an endeavour to reduce the hazards of its cultivation.

There is considerable variation between the various kinds and local stocks of field beans, and little has yet been done to improve the crop by plant breeding, although this is now receiving attention. The principal types are merely referred to as 'winter beans' and 'spring beans' of which there are several forms which can hardly be regarded as agricultural varieties in the strict sense of the term. There are many local commercial stocks of the common English winter bean, some of which are mixtures of several different types, but there are no named varieties which can be regarded as superior or suited to particular conditions. The common English spring bean, usually referred to as the English Horse bean, also exists in several local types, while the Tick bean is commonly a smaller seeded type. In the north of England and in Scotland the spring bean is known as the Scotch Horse bean, or more locally as the Carse and Kilbride beans. These English and Scottish types contribute the bulk of the bean crop grown in the United Kingdom, but an early type, the Mazagan bean, with a large flat seed, is also grown to a small extent.

The seed of the bean, which is the most important product of the crop, is usually fed to livestock, broken or ground into meal. Approximately one-half of the seed is carbohydrate and one-quarter protein, while the ash is relatively rich in potash and phosphorus. The nature of this concentrated food has made it a favourite for giving to working horses as a useful supplement to corn, but mixed with other foods it can be fed to practically any class of stock including milking cows, fattening sheep and bacon pigs. The value of the rest of the plant for feeding depends entirely on the management of the crop and in particular the time of cutting. Bean straw, when in good condition is similar in many ways to cereal straw as a feed, but has more protein. If the crop is cut before the stem blackens the straw is valuable, but in proportion to the lateness of cutting and damage due to weathering and mouldiness so does it become increasingly poor, and a considerable amount of bean straw from crops taken for seed is unfit for anything beyond trampling by bullocks in the yard.

The other crop belonging to the genus *Vicia* cultivated in this country is the vetch. There are several species of vetch cultivated for their seed or for forage in various parts of the world, but the one grown in this country is the species *sativa* which is utilised entirely as a forage plant, and is known as the Common Vetch or Tare. The plants are distinct from the bean in that instead of having a stiff, erect stem they cannot support themselves but rely on leaf tendrils to attach themselves to any convenient support. Tares are, of course, familiar as weeds, and it is only in comparatively recent agricultural history that these vigorously growing smothering plants have been utilised intensively in agriculture.

Although vetches are widely grown, and are suitable to a wide range of soil and climatic conditions in this country, their use as a pure crop is very restricted in the British Isles. The plant is extremely useful for mixing with cereals and other leguminous crops for cutting green or making into silage, and these are the principal uses to which it is put. Although doing best on calcareous soils of heavier nature, vetches are very adaptable and can be grown on poor soils, and in districts of low rainfall they can make much growth because of their relatively high drought resistance. The crop is hardy and can produce heavy yields of good forage where other leguminous plants are unsuitable, and the several forms of utilisation—cutting green, silage, hay or folding—give it a special value under suitable conditions.

The management and place in the rotation of vetches depends on the type grown, the method of utilisation and growing conditions. There are no agricultural varieties but there are spring and winter types, both annual in habit, but the latter quicker growing and more luxuriant. The Gore vetch is a bulky spring type which is grown to a small extent. It is not usual to take vetches in any fixed place in the rotation because of the varied ways in which they may be grown and utilised. When sown as a pure crop, it is common for them to follow a cereal, but they may also be taken as a catch crop. As a constituent of a silage mixture they are normally taken in the place of roots, but where winter sown their position varies. Vetches are not easy to make into hay, and this practice should be avoided in districts of high rainfall.

The only other important leguminous seed plant, beside the bean, grown in this country is the pea. These two crops are the only source of protein concentrate of this type available to the British farmer, and it is unfortunate that both have certain drawbacks to their cultivation. Cultivated peas belong to the genus *Pisum*, all forms of which are weak-stemmed annual plants which climb by means of leaf tendrils like the true vetches. Two species are commonly recognised: *arvense*, which includes the purple and red-flowered forms which have also coloured seeds and are grown principally for stock feeding, and *sativum*, which includes all the white flowered and white or green seeded forms, which are largely used for culinary purposes. The *arvense* forms are commonly referred to as 'field peas' because they used to be the only agricultural peas cultivated in this country, the *sativum* forms being restricted to market gardens, small-holdings, allotments and private gardens. This distinction no longer holds as rigidly as it used to, because although the *arvense* forms are still only grown on the field scale, there is a relatively large and increasing acreage of *sativum* forms grown as agricultural crops, and therefore eligible to be called 'field peas'.

It is usually asserted that the field pea is a native of the Mediterranean, and particularly of Italy and some of the neighbouring islands. There are certainly wild, purple-flowered peas in these areas, and the region appears to be more probable than any other. The cultivation of peas in Europe dates at least from the Bronze Age, and it seems likely that they were contemporary with the bean in their earliest history as agricultural plants, and were brought to this country at the same period. As in the case of beans, the cultivation of peas used to occupy a very much more important position in this country than it does to-day, due to the great reduction in the consumption of dried peas and pea meal, which used to enter conspicuously into the dietary of a large proportion of the population.

The whole economy of pea growing in this country has altered during the twentieth century. With the considerable reduction in the acreage of the forage field peas, there has been a gradual increase in the cultivation of green peas for picking green, packeting and, in recent years, of peas for canning. Considerable amounts of peas for stock feed have been imported because the crop has become increasingly unpopular among growers in this country owing to the difficulties of harvesting in a changeable climate and with labour shortage. The pea crop, like the bean crop, is a 'chancy' crop which is subject to many hazards when grown for the mature seed, and farmers have preferred to buy their protein concentrates from imported feeding stuffs. The good market for green peas has, however, offered an inducement to growers looking for a cash crop, and there has consequently been an expansion in this direction.

Peas are similar to beans in their preference for the lower rainfall districts of this country, and most of the field crop is grown in eastern England. Owing to the difficulty of handling the crop in wet and late districts, peas are not found growing as far north as beans. On the other hand, peas will not withstand really dry conditions, and although they are suited to lighter soils than beans there must be a fair soil moisture supply. Good drainage and an adequate lime supply in the soil are essential, and good sheep and barley land is usually regarded as suitable. When peas are grown for picking green they can, of course, be successful under higher rainfall conditions than when grown for the ripe seed, but too much rain is not desirable.

There are no winter-hardy forms of peas available to growers, although there are some types which are described as 'winter peas'. Some of the round-seeded green pea types are said to be better suited to autumn sowing than the wrinkled types, but any severe frost or persistently cold weather in the winter is liable to kill or badly damage the plants. The crop is, therefore, essentially a spring-sown one, and it is a common practice to take it in between two cereals. The rotational position varies considerably, however, according to the kind of peas that are being grown and the system of farming. Culinary peas in special pea-growing areas may be taken after various crops, while forage peas are a safe crop on newly ploughed grassland because of their resistance to wireworm. In some areas peas are mixed with a cereal and the mixed crop either taken mature as a grain crop or else cut green for forage. In both cases a well-balanced stock feed can be obtained.

The principal types of stock-feed peas are the Maples and the Duns, both being *arvense* forms with coloured flowers and pigmented seeds. The Maples have speckled seeds of varying shades of brown or yellow, the shape, size and colour varying considerably in different stocks. The most popular and highly prized samples are round, of a bright colour and with a white 'eye' or hilum, the best imported stocks being carefully picked over and even polished for the British market. It is commonly believed that home-grown samples cannot compete with the best imported product because the climate of this country prevents the growing of a round seed with a good colour. This is not true, however, and if care is taken to grow good stocks there is no trouble about the seed characters. An improved high-yielding and round-seeded variety named Marathon Maple was put on the market a few years ago, and is at present probably the highest yielding Maple for good soils. It is, however, rather late maturing and the seed is a dark colour, with a black eye, and is therefore not popular in some districts. It should be pointed out that such superficial characters as seed shape and colour have no effect on the feeding value of the seed, and the only criterion should be yield and suitability for particular growing conditions. Where Maple peas are used for pigeon feeding, and where they have had a market for human consumption, considerable attention is paid to the appearance of the seed, and although there may be justification for the latter, there is none for the former.

There is considerable difference in the time of ripening between different stocks of Maples, and some attention has been given to the breeding of early ripening forms for late, and high rainfall districts. By hybridisation between native stocks of Maples and quick-growing foreign strains and green pea varieties, very early types have been obtained which are now under trial.

The Dun pea stocks show a similar wide range in time of maturity, seed characters and length of straw to the Maples, but as a group the Duns at present available are earlier. The seeds are a grey or dun colour, turning chocolate brown with keeping, as do some of the Maples. The popular types have well-shaped and round seeds with a white eye, although in some districts forms with black eyes are grown under the name of Black-Eyed Susans. Dun peas are used almost exclusively for stock feed either split or ground, although not many years ago there was a small export trade of these peas to the continent.

Whereas the total acreage of peas for stock feeding in the United Kingdom was only 37,000 acres in 1939, the area of green peas in England and Wales alone in the same year was five times as great. Virtually the whole production of both classes of peas is concentrated in England, and the area devoted to green peas picked green has in recent years exceeded that of the peas for stock feed, while the acreage of dried green peas has been nearly double that of the stock feed acreage. Peas for canning have been the smallest contributors to the total pea crop and have been grown on only a quarter of the acreage devoted to peas picked green for market. Although these figures and proportions have shown considerable changes in recent years, they show fairly accurately the position of pea growing in this country.

The largest acreage of green peas is devoted to those for packeting, and of these the most extensively grown is Harrison's Glory. This pea, and the small round type known as Prussian Blue, is also used for canning after being first cooked. The round-seeded Prussian Blue used to be the most widely grown pea for drying and packeting, but with the introduction of the wrinkled-seeded 'marrow fat' Harrison's Glory, this latter type has become the most important variety for this purpose. There are other similar types, such as Koopman's Glory, and most of the stocks are of continental, particularly Dutch, origin.

The canning of fresh green peas is a recent commercial development in this country, and the industry promises to offer a useful addition to the cash crops of the farmer. Special varieties of peas are used for this, because it has been found that only some will answer the requirements of retaining their shape, colour, flavour, texture and be free from splitting. The chief worry of the canners is to extend the season as far as possible and avoid peaks in delivery. For this purpose every effort is made to plan a succession in the field by using varieties of different times of maturation. Common canning varieties are Thomas Laxton, The Lincoln, Delicatesse, Admiral and Senator.

The varieties of green peas for picking green and marketing fresh are so numerous and there is such considerable confusion in their identification and naming that it is more profitable to consider only the main types or groups. From the grower's point of view the important consideration is the season for picking, and on this basis varieties can be considered as early, midseason or late. The length of straw or haulm is a very variable character between varieties, but they are usually considered conveniently as short, medium and tall. Although, when the peas are picked in their best condition for eating, the mature characters of the seed cannot properly be distinguished, these mature characters are used to characterise and group varieties as round, dimpled or wrinkled. The actual choice of a variety for growing on the farm will depend on these characters as well as others such as yielding capacity; the relative weight of peas and pods; the resistance to disease and pod blemishes; the colour, size and shape of pod; the shelling characters; and the flavour, colour, shape and size of seeds.

The economic value of the green pea crop is primarily a matter of the feeding and nutritional value of the seeds. When mature, green peas have essentially the same composition as the stock-feeding *arvense* types, and are a concentrated protein and carbohydrate food. Peas picked green have a higher percentage of water, and a considerable proportion of the carbohydrate occurs as sugar. The actual composition of the peas depends on the stage of growth at which the pods are picked and the length of time and conditions of storage. Young peas are the most palatable and digestible, while long periods of storage under warm conditions causes loss of nutriment in the seeds.

Attempts have been made from time to time in this country to introduce other seed legumes for cultivation on an agricultural scale. The particular types of French beans known as haricots have been tried, and attention has been devoted to the special form which is used for canning. The climate of this country is, however, not suited to complete maturing of these plants which belong to the species *vulgaris* of the genus *Phaseolus*. It is, however, possible to grow the French bean and the related runner bean (species *multiflorus*) for their green and edible pods, and some thousands of acres are cultivated in England and Wales.

The climate of the country has also prevented the successful introduction of the soya bean (*Glycine soya*), which is unique among leguminous crops in the extremely valuable seed which it bears. This seed may have over a third of its weight as protein and a fifth of its weight as oil, with the rest of its food reserve as carbohydrate. This composition, combined with the relatively high digestibility of the seeds or the meal from them, make soya beans an outstanding food crop, but economic cultivation in this country is at the moment out of the question.

Considerably greater promise of economic cultivation is offered by another leguminous plant which has been cultivated on a very small scale in this country for some years—the lupin. This plant, which is a native of Mediterranean countries, has been studied intensively in recent years with a view to producing improved types for agricultural use. Two species *Lupinus luteus*, a yellow-flowered form, and the purple-flowered *L. pilosus*, have shown most promise. They are both annuals and they may be cultivated for their seed or their herbage, the seed, like that of the soya bean, having a very high percentage of protein, the biological value of which is said to be better than in most legume seeds. As a green forage, the crop has the special virtue of remaining remarkably succulent for a long time.

Great impetus to the cultivation of the lupin has resulted from the development of the so-called 'sweet lupins' which are free from the alkaloid which made the crop risky to feed to animals. This great improvement, combined with the selection of forms which do not so readily split their pods, and therefore which do not shed the seed in the field, has caused renewed interest in the crop. The plant has the particular virtues of being able to withstand a greater degree of soil acidity than other leguminous plants, and of growing well in dry climates provided there is a reasonable water supply in the soil. It is not a crop for acid or very acid conditions as is commonly supposed, nor will it succeed on wet, very heavy or very chalky soils. It is being advocated for lamb fattening as a superior substitute for rape, but up to the present has not received much attention from agriculturalists in this country. Its value depends on its capacity to produce good yields of seed, and also on its success as an annual green forage crop under conditions which are unsuited to other legumes.

Chapter XIII

LEGUMINOUS PLANTS: HERBAGE TYPES

Wherever there is cultivated grassland in this country there is at least one species of leguminous herbage plant available for mixing with the grasses to form a sward for grazing or cutting. These herbage legumes are an extremely important component of grassland husbandry and they play a vital part in the maintenance of an intensive livestock husbandry at a high level of productivity. Grassland in temperate regions has the great virtue that there are suitable leguminous herbage plants that blend well with turf-forming grass species to produce an excellent herbage and green forage for grazing animals. This blending of leguminous plant and grass not only provides a stock feed of the first importance for maintaining animals through the growing season, but the legume has special virtues in increasing the productivity of the sward, in supplying keep when the grasses may be making little growth, and in helping to build up the fertility of the land. The agricultural value of grassland is dependent not only on the particular species of grasses which dominate the sward, but also on the amount and the particular leguminous species which are growing with the grasses.

The leguminous herbage plants are an essential feature and a necessary requirement of the highest expression of rotational farming and alternate husbandry with their dependence on the temporary ley for the maintenance of fertility and the intensive feeding of livestock on home-grown produce. These plants are important also in providing suitable crops which can be grown in a pure stand in arable rotation, while their contributions to permanent swards, 'natural' grassland, and out-run grazings, are of the first consideration. The improvement of poor, neglected, or derelict grasslands is commonly sought by introducing some suitable leguminous herbage plant; standard seeds mixtures for agricultural grassland for practically all circumstances contain their proportion of a legume; and grassland management is almost always directed to the maintenance of a suitable balance of grasses and leguminous plants for the types of sward required. Probably more has been done by the making available of suitable and improved forms of leguminous herbage plants in the raising of the standard of grassland productivity than by any other means.

There are several distinct botanical genera and species, and many different agricultural varieties, of leguminous herbage plants that can be grown successfully in the British Isles. Most of them are natives of temperate cool regions, many are introduced from other countries, while some appear to be indigenous to this country. Over wide areas of the country the climatic and soil conditions are very well suited to the economic cultivation of some leguminous plant for its green herbage, while in most parts there is a choice of more than one kind available to growers. There is considerable variability in the growth type between some of the species, and the appropriate form may be chosen for the particular condition, management and utilisation. Some are long-lived perennials, others persist only for short periods, and a few are annuals. There are forms that are pre-eminently suited to grazing, forms which may be grazed or cut, and forms which are essentially best adapted to cutting and using in the fresh green state, or made into hay or silage. Although all are most successful on non-acid soils, there are a few which will tolerate more acidity than others, and some which are essentially chalk plants. A considerable degree of drought-resistance characterises some of the species, and the perennial forms are all sufficiently frost-hardy to withstand the winters generally experienced in this country.

The nutritive value of leguminous herbage plants, as in the case of grasses, depends largely on the stage of growth when grazed or cut, whether fed green or made into hay or silage, and the efficiency with which the conserved fodders are made. When used as cut forage, leguminous plants commonly have a higher percentage of digestible protein and of calcium than does cut-grass forage used in the same way, but some leguminous plants are apt to become more fibrous than the better grasses as they mature. The relative agricultural value of the different leguminous species is primarily a matter of the yield of stock food each is capable of giving, their suitability to different forms of management, and their range of adaptability to growing conditions. Some species are consistently high yielders over a wide range of soil and climatic conditions as experienced in this country, while others show distinct preference for a limited and particular environment, and consequently are not suited to general cultivation. The yielding capacity is naturally largely a reflection of the size and vigour of growth of the particular species, some being so small and relatively unproductive that their chief value lies in their capacity to persist under conditions where no other leguminous plant could survive usefully.

The true clovers, which belong to the genus *Trifolium*, are easily the most important leguminous herbage plants of agricultural land in this country. There are some seventeen species of clover which are considered to be native to the British Isles, but only three are of any great agricultural value. The forms of these species used by the farmer do not always correspond with those found growing wild, but have been developed by selection and hybridisation. Some of these cultivated forms have been introduced from other countries, while a distinct species not occurring as a native was brought from Sweden during the last century and has found a useful place in British agriculture. By far the most important clover species used agriculturally are the red clovers (*Trifolium pratense*) and the white clovers (*T. repens*), the former being the commonest legume used in the formation of short-term temporary swards, and the latter for temporary swards of longer duration and permanent swards.

The red-clover species is very variable and wild forms are found growing on permanent and uncultivated grasslands in many parts of the country. Wild red clover, which is usually considered to be a distinct subspecies (*spontaneum*), is not a valuable agricultural plant, although it is an early grower in the spring and is very persistent and long-lived once it establishes itself. This form is never sown because its yield and contribution to the keep for livestock are insignificant. It is a low-growing, stemmy plant which seldom occurs in any profusion in grassland, and the size and habit of growth of the plant, although variable, do not make it a particularly suitable plant for either cutting

PLATES

19-27

OPENING THE POTATO CLAMP IN WINTER

PLATE 19

The potato is an outstandingly important food crop for human consumption and is grown extensively throughout the British Isles, although
the important potato growing districts are localised in a few areas. The
plant is the only representative of the genus *Solanum* cultivated as an
agricultural crop in this country, and its great value lies in the large
amount of food which is stored in the tubers below ground. The potato
crop can produce more than twice the amount of human food to the
acre than wheat, and has the special virtue of being easily prepared for
the table. This country can grow sufficient potatoes to meet all normal
demands, but there is a certain amount of importation of early potatoes
at the beginning of the season. Most of the home-grown crop is clamped
and marketed through the winter as the demand arises, the varieties
used being the slower maturing 'main-crop' types.

SOWING SUGAR BEET

PLATE 20

Sugar beet is both a cash crop and cleaning crop, and preparatory cultivations before sowing are carefully carried out by good growers. After deep ploughing, the soil is worked down to a fine and firm tilth, and artificial fertilisers applied before drilling the seed. Sugar beet is very susceptible to poor soil conditions, and successful germination and establishment can only be expected under good physical conditions of the soil. The preparation of the land for sugar beet and other cleaning crops is an important feature of crop rotations in Britain, the soil cultivations performing an essential function in helping to keep weeds under control.

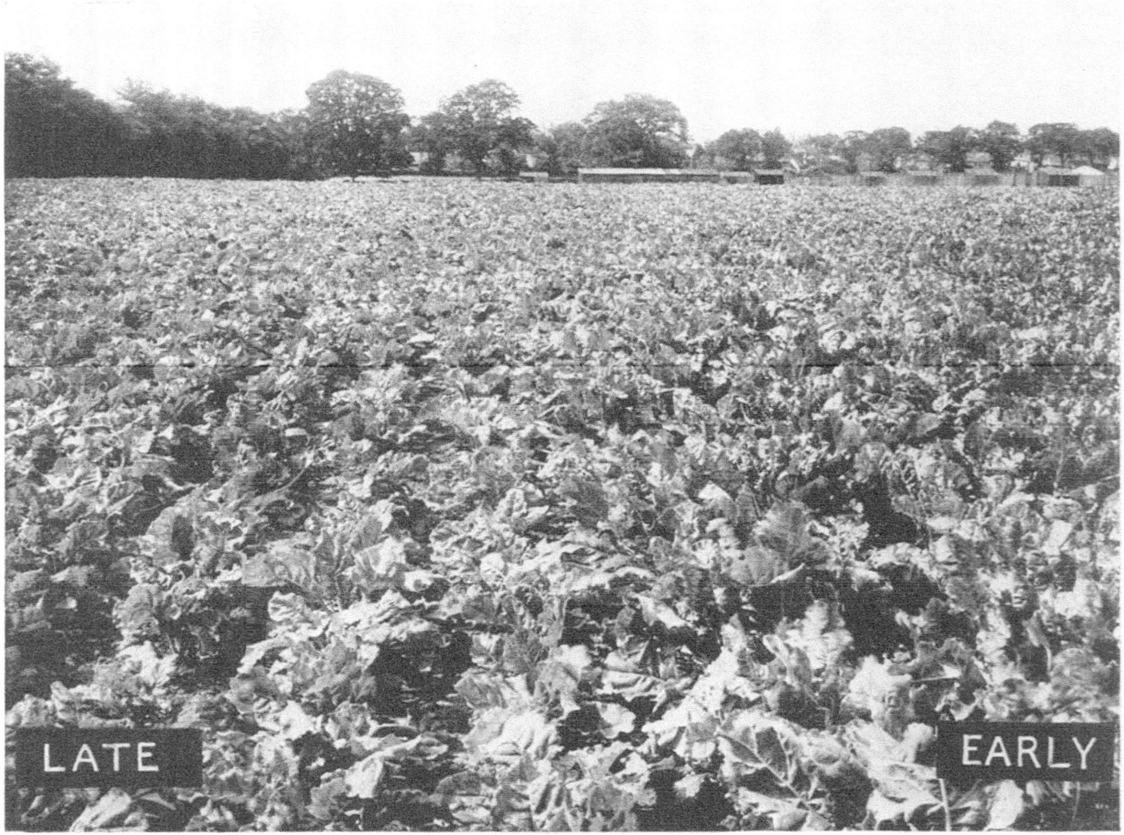
LATE
EARLY

PLATE 21

Sugar beet is the only agricultural crop of temperate regions that is grown for commercial sugar production. The plant belongs to the genus *Beta*, which is important agriculturally only because it includes the beet and the mangold. The sugar is stored in the swollen tap root of the beet, and the amount of sugar produced to the acre depends in the first place on the weight of roots and the sugar percentage of the root juice. Given suitable conditions, the crop grows very rapidly, but for heavy yields of sugar there must be a sufficiently long growing season and adequate sunshine. The crop often suffers in this country from late sowing, and considerable differences in yield may result, even in the same field, from differences in sowing time. A late-sown crop may remain under-developed throughout the whole growing season as may be seen from the development of the foliage, while the root weight and sugar percentage suffer correspondingly.

PLATE 22

Mangolds are very closely related botanically to sugar beet, and are usually considered as belonging to the same species. Mangolds are, however, not grown for the extraction of sugar, but are cultivated as a succulent fodder for livestock to be fed in the winter. The crop is normally clamped after lifting, and the most popular types are those that grow well out of the soil and are easy to lift. There are many strains differing in root shape, size, dry matter content and keeping quality, and the heaviest root yielders are by no means the most economic to grow. Mangolds are suited to the drier and sunnier parts of the country, but they can only be grown successfully under such conditions on deep and fertile soils.

FIELD BEANS

PLATE 23

The field bean is grown as a pure crop for the mature seed, but it is also included in silage mixtures and is then cut green and the whole plant fed to livestock. The field bean and the field pea are the only leguminous plants cultivated agriculturally on an appreciable scale in Britain for the mature seed which is used as a concentrated feeding stuff. Both crops are very 'catchy', and the yield fluctuates widely from season to season, so that neither is a popular crop with the farmer although the seed is extremely valuable as a source of protein. Beans are cultivated in many arable districts in Britain, but the greatest acreage is on the heavier classes of soils in the midland and eastern counties of England.

PULLING FLAX

PLATE 24

Flax or linseed, is the only cultivated plant belonging to the genus *Linum* which is grown on an agricultural scale in Britain. The same species contributes the fibre-producing types and the seed-producing types, but different varieties are used for the two purposes. Flax for fibre is a more suitable crop for the climate of this country than is linseed because it requires moist and cool conditions. The fibre, which is among the most valuable of all plant fibres, occurs in the stem and is used for manufacturing linen and other fabrics. The crop is usually pulled, and special machines have been designed for the purpose, although in some parts the pulling is done by hand, which is a more satisfactory method of dealing with a weedy crop.

HARVESTING CELERY

PLATE 25

Celery is a vegetable crop for intensive cultivation, and is grown on a field scale on rich fen and skirt soils. The marketable product consists of a heart formed by the swollen leaf stalks which are blanched by moulding the soil around the plants. The family Umbelliferae to which the celery belongs is represented by a large number of wild plants in Britain, but the only other cultivated plants of agricultural importance in the family are the carrot and the parsnip, both of which are valued for their swollen tap roots.

CUTTING MAIZE FOR GREEN SOILING

PLATE 26

Many attempts have been made to introduce maize as an agricultural crop into this country, but the climate has so far prevented the full exploitation of the plant which requires a combination of warmth, sunshine and moisture not commonly met with in Britain. Maize is a cereal of sub-tropical origin, and is valued particularly for the grain which has a relatively high oil content and is prized as a livestock feed. In this country it is difficult to obtain good yields of grain, but some vigorously growing varieties are grown for cutting green and either feeding direct or making into silage.

COUNTING SEEDS FOR GERMINATION TEST

PLATE 27

Seed testing ensures that the farmer can obtain seed of good quality with regard to purity and germination capacity. Testing is done under controlled conditions and special methods are used for individual crops. Samples of seed for the germination test are obtained quickly and conveniently by means of special equipment which by suction enables the worker to count out 100 seeds evenly spaced. These seeds are then placed on the appropriate germinator, wet filter paper being a common medium for the test. (See also Plate 28.)

or grazing. Because of its general growth characters it is only of any value in permanent swards, and although it commonly establishes itself on old grass, it is not deliberately encouraged by the farmer.

Of infinitely greater agricultural value are the so-called cultivated red clovers, which in all probability arose from some form of wild red, but whose origin is not known with any certainty. These cultivated red types are grown throughout Europe, and were first introduced into this country from the Low Countries in the eighteenth century and were responsible, with turnips, for the development of the rotational system of farming. Since the early days of the introduction of red clover into this country a large number of strains have been developed, and between them have come to occupy the most important place as the most widely used herbage legumes for temporary leys.

The habit of growth and persistency of the various types of red clover differ considerably between the different strains, but their characteristic quick and vigorous growth from seeding, and their relatively large production of nutritious vegetative parts make them eminently suitable for cutting green and making into hay. They all grow with a tufted, non-creeping habit forming a cushion or rosette of leaves in the early part of the season, and later sending up several branching and leafy flowering stems which are the valuable and preponderant contribution on cutting. The basal rosettes may be grazed, and the various strains differ in their suitability and capacity to withstand the action of the grazing animal.

The agriculturalist recognises two main types of cultivated red clover according to the way in which they may be managed, and the botanist distinguishes these two types by their rhythm of growth and certain characteristic features in appearance which are considered to be sufficiently distinct to establish them as subspecies. The first of these subspecies is *praecox* and includes all the forms called Broad Red, Early Flowering or Double Cut Clovers. The strains in this subspecies are more or less biennial and do not persist as long as the strains in the other subspecies; they are on the whole early flowering and are capable of flowering twice in one season when they are cut back at the first flowering.

They are used essentially for one-year leys and show rather a high degree of susceptibility to 'clover sickness'. The second subspecies, named *serotinum*, includes the agricultural types known as the Late Flowering or Single Cut Clovers. These persist longer, usually up to four years, and flower only once in the season and then later than the Early Flowering Clovers. There is, however, no distinct and fixed difference in flowering date between the two groups because the strains differ in this character, there being late Earlies and early Lates, which with the early Earlies and the late Lates, give a continuous range of flowering time.

The important agricultural and botanical characters distinguishing the two main types and various strains of red clover are the time of flowering, habit of growth, vigour of growth from seeding and at different times of the year, persistency and yield. The time of flowering is important because it affects most of the other characters, and particularly the yield, proportion of hay to aftermath, contribution in the seeding year and persistency. As a general rule the Earlies yield more than the Lates in the seeding year, while the Lates outyield the Earlies from the first harvest year onwards in the hay, but not in the aftermath. The Earlies tend to have a more open growth, are more erect, and produce fewer flowering stems than the Lates, which, however, show considerable differences between the strains in this character of 'tiller' production.

There are four different kinds of strains available in both the Early and Late groups, according to their origin—foreign commercial, home-grown commercial, local 'indigenous' and bred strains. American Medium, Chilian and Britanny are examples of foreign commercial Early strains; and English Broad Red is the standard home-grown commercial Early strain, with its local Norfolk, Suffolk, Cotswold, Berkshire and other county variants; Vale of Clwyd and Dorset Marl are old established local 'indigenous' strains, and S 151 is a new-bred strain which, it is claimed, is more persistent than ordinary Broad Red. Among the Late Flowering forms there are the Swedish and Danish foreign commercials; the widely cultivated English Late Flowering with its best-known Cotswold and East Anglian variants;

and the Cornish Marl and Montgomery (of which there are two forms) which are similar in being profusely branched and close growing types and are typical of the local 'indigenous' strains. Recently the bred strain S 123 has been introduced as a dense growing strain suitable for hay and grazing.

There is no other herbage legume that can adequately fill the place of red clover in British agriculture, and it has no comparable alternative or competitor as the most important hay legume for short-term temporary leys. White clover (*Trifolium repens*), the only other species in general cultivation, has quite a different place in grassland husbandry and is in no way an alternative to red clover. It has been said of white clover that the realisation of its great importance, and the making available of cheap seed of good strains, have been responsible for as significant an agricultural revolution in modern times as the introduction of roots and red clover was in the eighteenth century. Although white clover can be regarded as a native of these islands, and is widely distributed wherever there is well-tended agricultural grassland of a permanent nature, its great possibilities and significance have only been fully realised in comparatively recent times.

The particular virtue of white clover is its suitability as a grazing plant. It is a perennial that creeps along the surface of the soil by means of runners that root at intervals and bear erect growing leaves and short inflorescence stems. It is capable of colonising new ground by means of this habit of growth and it mixes and blends with the best pasture grasses in a most successful manner, but it is very susceptible to the competition and shading of tall growing grasses. It is drought resisting because of the deep tap root of the plants establishing from seed, but under really dry conditions its growth by runners is restricted. Like all clovers, white clover will not grow well where the soil is acid, but it can succeed on a wide range of soil types and is stimulated by phosphatic minerals. There is no other clover that thrives so obviously under controlled grazing by all classes of stock, and its adequate representation in grazing swards is one indication of a good pasture. Poor pastures can be improved by the introduction of white clover, and the value of a ploughed-out sward for putting through an arable rotation is considerably enhanced by the presence of this plant.

White clover is an extremely variable plant and the botanical species is represented by three subspecies. The first, and least important to the agriculture of this country, is the giant or Ladino Clover (*giganteum*) which because of its size can be used for cutting green when mixed with a bulky grass species. Ladino Clover differs from the other white clovers in its unsuitability for close grazing, while it can be grown successfully in hot, dry conditions, provided the soil is highly fertile and there is a higher lime status than is suitable for the common forms of white clover cultivated in this country. At the present time there is virtually no Ladino Clover grown in the British Isles, and it has never occupied a position of any importance, although it might be considered as a suitable legume for the driest parts of eastern England.

The two common subspecies which contribute virtually the whole of the white clover cultivated in Britain are the cultivated, Dutch, or 'ordinary' white clover (subspecies *cultum* or *hollandicum*), and the true wild white clover (subspecies *sylvestre*). The Dutch or cultivated white clover is a shorter-lived and more robust type than the true wild white clover, and is most useful for temporary grazing leys. There are various forms of this clover on the market including the English, common Dutch, Danish Mørso, and the various New Zealand stocks of the grades known as Pedigree, Certified Mother and Certified Permanent Pasture. The commercial stocks of English and Dutch white are usually the shortest lived and are thus suitable only for the more temporary leys, while the New Zealand stocks can be kept in economic productivity for a longer time. The comparatively recently introduced bred strain, S 100, is a large type with a long growing season, and is probably the heaviest yielding white clover available for temporary leys. The true value of S 100 has not yet been fully determined, but it promises to be the most valuable form of white clover cultivated in this country because of its vigour and capacity to remain in a high state of productivity for a greater number of years than the other stocks available, including the New Zealand whites.

For long-term temporary grazing swards and for permanent pastures, the true wild white clovers are commonly used. These clovers are valued for their persistency and more or less permanent nature under good management and suitable conditions. The seed of these clovers is harvested only from swards which have been down for ten years or longer, and there is an official certification scheme for their marketing. In addition to the stocks known simply as English Wild White, there are the local strains such as Kentish and Cotswold, and the bred strain S 184 which is more uniform than the commercial stocks. All the true wild white clovers are small growing, extensively creeping types with slow establishment from seeding and a rather restricted period of growth. They only thrive with the smaller grasses and under close grazing, and their general characters are such as to warrant their use only for grazing swards that are being treated as permanent pasture.

Of the remaining clover species used agriculturally in this country, though on a much more restricted scale than the red and white clovers, the most important are Alsike (*Trifolium hybridum*) and crimson clover (*T. incarnatum*). Alsike has a pinkish white flower head not unlike that of a large white clover, and is a small, tufted perennial plant introduced from Sweden. Its chief value is an insurance against the failure of late flowering red clover because it is hardier and less liable to 'clover sickness' than the latter. It is usually included in mixtures for temporary swards with the red clover, but it has not the yielding capacity and cannot be regarded as a real substitute for the reds, although it is certainly more suited to cutting than grazing. Crimson clover, which is a conspicuous annual plant with its elongated crimson inflorescences, is limited in its successful cultivation to the south and south-east of England where it is considered to be native to the lighter and warmer soils. It is a very quick growing plant, although there are varieties which differ in the flowering time. It is immune to 'clover sickness', and may be used in a one-year seeds mixture with Italian rye grass, or it may be sown alone. It is sometimes treated as a catch crop, and may be sown on autumn stubbles for feeding off the following year. In mild and sunny climates, and on suitable soils, crimson clover can be used for hay, silage or green manure, but its successful growth is very much dependent on the right conditions. There is, peculiarly, a white-flowered form which flowers about the same time as the late-flowering crimson form, but it is seldom used in this country.

On dry soils are often found small, yellow-flowered clovers contributing a certain amount of keep for grazing animals. The two common species are the Yellow Suckling Clover (*T. dubium*) and the hop trefoil (*T. procumbens*), both of which are annuals, the former chiefly on non-calcareous soils, and the latter preferring chalks or limestone. Neither of these plants has any great agricultural value except under very special circumstances, and they are very seldom included in seeds mixtures, being low yielding and only persisting for one year except by self-seeding. Other species such as the annual Subterranean Clover (*T. subterraneum*) and the perennial Zig-Zag Clover (*T. medium*) have from time to time been tried in this country, but neither has succeeded in establishing itself as a useful agricultural plant.

The true clovers rank as easily the most extensively cultivated herbage legumes in Britain because of the considerable range of botanical forms which offer themselves as useful plants for a wide range of conditions and different methods of agricultural management. The clovers are as essential to intensive grassland and stock husbandry in this country as are the more important arable crops to rotational farming, and their use in helping to maintain soil fertility in many farming systems is considerable. The only other botanical genera which contribute any plants of real importance to British agriculture are *Medicago* and *Onobrychis*, each of which has one species of particular agricultural value.

The genus *Medicago*, which includes some half-dozen wild native species commonly called 'medicks', is chiefly important because among its species is the cultivated lucerne, a plant not counted as indigenous to this country. The only native species which has achieved any importance, and this is very limited, is the Black Medick or trefoil (*Medicago lupulina*), an annual plant which cannot compare in usefulness with lucerne, although the perennial sickle medick (*M. falcata*) has a significance of its own because of its alleged participation in the parentage of some forms of lucerne.

The old established and 'pure' form of lucerne is a purple-flowered perennial belonging to the species *M. sativa*, which is a native of western Asia, Anatolia, mountainous Caucasia and north-west Persia. It is really of southern origin, therefore, and enjoys a true continental climate with hot summers and low rainfall more than the cool, high rainfall conditions of northern maritime countries. Lucerne was cultivated in its original home long before the Christian era, it was known in Greece, Italy and North Africa, and has since spread throughout Europe as far north as Scandinavia as well as migrating eastward into India and Asia. It is said to have reached England in the middle of the seventeenth century, to have been taken to South America in the eighteenth century, and North America in the first part of the nineteenth century. The crop is now world-wide, with the U.S.A. and the Argentine as the greatest cultivators, although it must be stated that some of these forms are not true *M. sativa*, but belong to the species *M. media*, which is said to be a hybrid between *sativa* and the sickle medick, *falcata*.

Lucerne is commonly regarded as the most valuable leguminous herbage or forage plant in the world, and it is true that for extent of cultivation and usefulness in areas where it grows successfully there is nothing to compare with it among the leguminous crops. Although its position in Britain is very much inferior to the clovers, lucerne is a plant which will repay considerable attention because of its obviously great attributes. The crop has never been cultivated on the scale that it deserves in this country, and in spite of its two-hundred years' history in British agriculture, the acreage has remained very small.

The true *sativa* form of lucerne, which until recently has been the type usually cultivated in this country, is an erect growing bushy herbaceous plant which develops a crown of leafy shoots from a well-developed tap root. Actually the shoots are borne on a short branched stem at, or near, the surface of the soil, and when the older stems grow up and flower, new buds are developed at the base. These buds are stimulated to develop when the older parts are cut or grazed, and a new crown of leafy shoots replaces the old one with a consequent renewal of the leafy parts. Provided the plant is given sufficient time to recover from its defoliation, it will produce more than one new crown in the growing season, and in this way provides growth for a succession of cuts or for grazing. Some types produce the new buds at or above the surface of the soil, and are thus suited primarily to cutting, while others develop their new buds below the soil on short, slightly creeping stems, a habit of growth which makes them suited to grazing. This habit of the plant is of particular importance with regard to the management of the crop, because the yield, the number of times of cutting, and the persistency of the plants, depend on the severity of the defoliation and the time allowed for recovery.

Lucerne has always been regarded primarily as a crop for dry conditions, and its great drought-resisting characters are chiefly responsible for its world-wide cultivation in areas with hot summers and quick growing seasons. In this country it has been grown chiefly in the low rainfall areas, but it has considerable climatic and soil adaptability, although it is not a crop for extreme conditions of soil poverty and dryness or for waterlogged and cold areas. It will not stand acidity, and thrives on fertile retentive soils with a good organic-matter content. The plant is rather slow to reach its maximum productivity, usually requiring about three years, but once well established it will maintain itself at an economic level for seven or eight years from the time of seeding if managed properly and if the growing conditions are suitable.

The commonest method of cultivation is to sow the crop by itself as a 'lucerne ley' and to cut it green as required, or make it into hay or silage. It has become a popular practice in recent years to mix a grass with the lucerne, primarily to help suppress weeds which are usually the chief reason for the limitation of the life of the ley. This practice has now developed to the stage when a balanced mixture of lucerne with one or two grasses is used to establish a temporary ley which can be used for cutting or grazing. With the development of types more suited to grazing this procedure offers wider scope for the successful utilisation of the plant, because it mixes and blends extremely well with vigorous grasses such as cocksfoot. But any extended use of lucerne is dependent on a better understanding by the grower of the requirements, limitations and possibilities of the plant, and one of the reasons for its limited cultivation is a lack of knowledge and appreciation in this direction. Potential growers are often deterred by the thought of having to inoculate the seed with the appropriate strain of nodule-forming organism, but this is a simple matter. Mismanagement in the form of sowing on a weedy seed bed, cutting or grazing too drastically in the early years, failure to allow adequate time for recovery between cuts, and lack of appreciation of the best conditions for the growth of the crop in order that it can overwinter successfully, are the chief causes for poor results with the crop.

The older varieties or strains of lucerne were in reality groups of geographical races named after the country or region of origin, such as Provence, German, Turkestan and Arabian. Some of these were distinct in their botanical characters, the Common or Provence form, for example, having a low-set crown and ascending branches. In this country, Hungarian and Provence strains have proved very successful, and the Provence types which have resulted from seed taken from English grown crops have been the most widely grown and the best liked. In recent years, the American variety called Grimm has been sown more than any other, largely because of the war, and it has shown itself to be very satisfactory. Grimm is considered to be a hybrid between the species *sativa* and *falcata*, and therefore as belonging to the species *media*, which also includes such varieties as Saskatchewan, American Variegated and German Variegated. The Canadian varieties, Ontario Variegated and Rainy River probably have a similar origin, while the New Zealand variety Marlborough can be recommended as a suitable persistent type.

The choice of strain is very important in lucerne, and the origin of the seed of a named strain or variety needs careful attention on the part of the grower. The recent stocks of Grimm from the U.S.A. and Ontario Variegated and Rainy River from Canada have proved to be the best in trials during recent years, while in earlier trials Provence was slightly the better. English grown seed of lucerne is very variable in quality because of the uncertainty of obtaining a well-ripened sample, but providing this has been obtained there is no reason to suppose that home-grown stocks are inferior to imported. It should be remembered, though, that the climatic conditions of this country appear to be practically at the limit of successful lucerne cultivation, and the effect of repeated taking for seed may eliminate certain valuable types from within a strain, while at the same time selecting types which seed well under the seed-growing conditions.

The other species of *Medicago*, black medick or trefoil, is of much less agricultural value, and has little in common with lucerne except certain botanical characters. It is a quick-growing annual with a trailing branched habit and little heads of yellow flowers resembling certain yellow clovers. Although widely distributed in Britain, black medick prefers limestone and chalky soils, and it is in such areas that it is commonly found as a cultivated plant. It is sometimes sown alone and cut as wanted for feeding green, or else undersown in cereals for grazing. Seeds mixtures for chalky soils often include black medick, which, with Italian rye grass, supplies useful early spring grazing. There are no varieties recognised or named, and the crop is not variable botanically, a fact which together with its limited agricultural value has meant that it has not received any serious attention from plant breeders.

From the agricultural and certain botanical points of view, the last of the more important cultivated leguminous herbage plants, sainfoin (*Onobrychis sativa*), has some characters in common with lucerne. Sainfoin is a perennial, deep rooting and drought resistant, forming at first a crown of leaves from which arise the usually ascending flowering stems. It does not form the same sort of multiple crown as does lucerne, the leaves have many leaflets borne in pairs instead of the three of lucerne, and the flowers are rosy pink. Like lucerne the crop can be used in more than one way—cut and fed green, made into hay, or grazed: it may be sown alone as a 'sainfoin ley', or it may be used in seeds mixtures. Sainfoin has not such a wide range of adaptability as lucerne, and is essentially a chalk-loving plant which thrives best in the drier areas and with reasonable sunshine. The plant is found growing wild in some parts of the country, and is considered to be a native, but the common name is obviously derived from the French and means 'healthy hay' on account of its reputation for being suitable for animals not in a thriving condition.

There are two distinct types of sainfoin—Common and Giant —the former usually considered as the older and the original type cultivated in this country. Common sainfoin is longer lived than Giant, although usually only kept in production for three or four years, is lower growing and flowers normally only once in the growing season. Giant sainfoin, which is sometimes considered as a distinct botanical variety or subspecies, is a more vigorous and bulky grower, flowers twice in the season, and usually only persists economically for two, or perhaps three years. Both the Common and Giant forms have Certification Schemes for the recognition of local strains or types which are distinct in appearance and behaviour, and particularly suited to the farming conditions in their own areas. There is more diversity among the Common sainfoins than among the Giants, and four types—Cambridgeshire, Cotswold, Glamorgan and Hampshire—are eligible for certification in the former group, and only two—Dorset and East Anglia—in the latter. There are, in addition, many local named stocks which are recognised in certain areas, and which are of reputable performance.

The clovers, the medicks and sainfoin comprise all the important herbage legumes used agriculturally in this country. There are some native species, such as the bird's foot trefoil (*Lotus corniculatus*) and kidney vetch (*Anthyllis vulneraria*), and the introduced sweet clover, or Bokhara clover (*Melilotus alba*), which have a certain agricultural use. The bird's foot trefoil is a small perennial plant widely distributed in permanent grassland where it contributes some useful grazing in the absence of other legumes, but it is rarely used in sown swards. Kidney vetch is a very drought resistant perennial that is only found in any amount on chalk soils where it has a useful part to play as the most successful legume of any bulk that will remain in a state of productivity on dry chalk soils in times of drought. Both of these plants are good examples of native species which have never quite reached the status of being standard cultivated agricultural plants, and their limited use and variability have not encouraged any attempts at their improvement. The sweet clover is a native of the drier parts of Europe and Asia. It is an annual or biennial plant with great powers of drought resistance, and is said to be very valuable as a soil fertility builder. It has the habit of lucerne, but is more stemmy and fibrous, and can be used in short leys or for cutting on poor dry soils. Sweet clover has received considerable attention as an agricultural plant in the U.S.A., and certain other countries where dry-land crops are important, but it is doubtful if there is much use for it in Britain owing to the availability of more economically valuable leguminous plants better suited to the agriculture of the country.

All the commonly grown leguminous herbage plants are cross-pollinated by insects, and, as in the case of the grasses, which, however, are wind-pollinated, special considerations are involved in breeding, maintenance of strains and seed production. Controlled pollination by bees is commonly used in breeding for self- or cross-pollinations during the strain building, which usually involves the selection of types from among the wide variation occurring naturally and the subsequent blending of similar types to ensure as great a degree of uniformity as is possible. Adequate isolation of nucleus stocks and field crops for seed is necessary, and the commercial handling of several strains of the same species needs careful control and strict supervision.

Chapter XIV

BRASSICAS

The cultivated plants of the family Cruciferae are more generally familiar than most agricultural crop plants because all the more important ones belong to the genus *Brassica*, which also includes some of the commonest and most widely grown garden vegetables. The family Cruciferae owes its name to the structure of the flower, in which the four petals form a symmetrical cross, and the uniformity of this flower type makes this 'Mustard Family', as it is sometimes called, easily recognisable. In addition to such familiar vegetables as swedes, turnips, cabbages and cauliflowers, cruciferous plants are well known by such popular garden flowers as wallflowers and stocks, and such ubiquitous weeds as shepherd's purse and charlock.

Cruciferous plants are widely distributed throughout the world, but the largest number are found in the temperate regions of southern Europe, Asia Minor and in China. They have long been recognised as plants possessing special characters which make them valuable as fresh vegetables and succulent fodder for livestock. They are all nitrogenous, and many are pungent, bitter to the taste, and possessed of antiscorbutic properties. The common scurvy grass, for example, is a medicinal plant with a long history, and horse-radish has served as a condiment for centuries.

But the family owes its extremely important agricultural position to the varied types of succulent fodder plants which are found in the one genus *Brassica*. This genus is unique in its variability, and there is no other group of cultivated agricultural plants of temperate regions which can offer such an extraordinary range of specialised habits of growth and edible parts. It is this feature which makes the Cruciferae third in the list of botanical families as supplying agriculture with a variety of cultivable plants which have come to occupy a position inferior only to the Gramineae and Leguminosae. In the early years of their history as cultivated plants, the brassicas were probably only grown as vegetables and potherbs, their use in this way being familiar in

Greece, Rome, Asia Minor, India and China. The extensive agricultural cultivation of these plants followed later, with the development of rotational farming and the practice of drilling crops in rows, while the growth of more intensive livestock husbandries added a further stimulant. But even so, the turnip was cultivated on a limited scale agriculturally in this country and Europe before the sixteenth century and consequently has a long history as a field crop.

The full extent of the great range of plant forms known as 'brassicas' cannot be appreciated from the forms which are cultivated in any one country. Nevertheless, the types met with in this country show a considerable diversity, although all cannot be regarded as important agricultural crops. The important economic characteristic of these plants is that they can produce large amounts of succulent fodder in their vegetative parts owing to the development of various organs and tissues to an extremely vigorous degree. Many brassicas are biennials, growing strongly in their first year and storing reserves of food in various parts such as stems, roots, leaves, buds and inflorescence branches, and running up to flower and setting seed in their second year. These types are grown for the product of the first year's growth only, the plants being utilised in various stages of development before they start to develop the second year's sexual growth. There are others, like the seed mustards and the coleseeds, which are annuals and which are grown for the seeds; these are not fodder plants but are valuable for particular storage products which have other commercial applications.

The cultivated forms of the genus *Brassica* are divided into a number of species and botanical groups which are not always easy to distinguish, and it is necessary to consider even the comparatively few forms in this country in relation to their botanical arrangement if they are to be properly understood. Some of these forms have only recently become agricultural crops of any importance, and they were formerly confined principally to market gardens and small holdings. It is probable, however, that in the future some will be taken more and more as farm crops as the demand for fresh vegetables increases, and the necessity for growing increased supplies of home-grown succulent feed for a more intensive livestock industry becomes necessary. The climate of the British Isles is particularly suited to the cultivation of such crops which require moist temperate conditions for the highest development of their vigorous vegetative growth.

The most variable *Brassica* species is *oleracea*, sometimes called the cabbage group because it includes the common cabbage and other closely related leafy and succulent stem types. In this group are found forms which develop single large and swollen buds or heads as in cabbages and savoys; numerous small swollen buds on a long stem as in brussels sprouts; swollen flower stems and inflorescence branches as in cauliflowers and broccoli; swollen main stems as in marrow-stem kale and kohlrabi; and branched or unbranched leafy heads as in the various kinds of kale. Every vegetative part of the plant except the root is represented by one or other of these types, and the edible portion may be used in some cases only as human food, in others only for livestock, while certain forms are used in both ways. As human food the product is usually cooked, but it may be pickled; the types used for stock feed are commonly fed green, either cut or folded, but some types can be made into silage.

It is commonly assumed, and usually stated, that all these cultivated plants of the 'cabbage tribe' are descended from the wild sea cabbage which is a native plant of the coasts of the Mediterranean, of west and south Europe, and England, and has the same botanical name of *Brassica oleracea*. The idea has the merit of simplicity, but it is hardly convincing, although it is true that some of the cultivated *oleracea* forms such as the branching kales are very similar to some forms of the wild cabbage. This wild brassica is certainly the nearest naturally occurring plant to the cultivated oleraceas, but as in other cases of 'wild progenitors' of cultivated plants, its relationship is still open to discussion.

Cauliflowers, broccoli, sprouting broccoli, brussels sprouts and special kales such as early kale or cottager's kale are all grown essentially for human consumption and are regarded as market-garden crops. Their total acreage is small and they do not really enter into the economy of British farming to any material extent except in certain specialised and localised areas which concentrate on vegetable growing. They are, however, plants which can be cultivated on a field scale, and considerable amounts of the more important ones like brussels sprouts, cauliflowers and broccoli are produced in this way. The production of these crops is essentially an intensive agricultural undertaking, requiring a considerable amount of hand labour particularly for the harvesting and preparation for the market of such perishable vegetables for human consumption.

The other brassica crops in this cabbage group, with the exception of the cabbage itself, which is in the nature of a dual purpose crop, are farm crops for stock feed only. The cabbage, which is grown for the market, or animal consumption on the farm, is a very accommodating plant in that there are a large number of forms suitable to varying methods of cultivation and different form of usage. There are varieties suited to sowing at different times of the year which correspondingly come for use at varying seasons, and, according to the time of planting and the choice of variety, it is possible to ensure a supply at practically any time of the year. Some varieties are grown only for human consumption, and others usually only for stock, but there is no absolute distinction. There are quick maturing forms for spring sowing and summer use, and slower growing, hardier varieties which can stand most winters. The shape of the 'heart' varies, and drumhead, flat, oxheart and conical types may be distinguished, while leaf colour, texture and character vary considerably with characteristic 'blistering' in the savoy forms. The great agricultural value of cabbages is due to their possible utilisation at practically any time of the year, their safe utilisation for all kinds of stock, their freedom from milk tainting, their possible use at any stage of growth or maturity, and their high yields of nutritious, easily digestible and palatable forage. The crop can be grown under virtually all farming conditions in this country, but it thrives best on fertile, retentive soils, and like all *oleracea* plants is a hungry feeder which responds to intensive management.

The agricultural kales may be used in the same ways as the agricultural cabbages, and may be cut or folded. The thin-stemmed leafy types vary to some extent in the amount of crown branches and leaves they produce, the typical thousand-head type producing its valuable product as a mass of leafy shoots. It is valued chiefly for its winter hardiness and the keep that it will provide in the late winter, but by suitable adjustment of sowings and time of utilisation it can be given a longer season. The marrow-stem kale has a considerably more succulent and bulky stem and thicker leaf stalks than the thousand-head type, and these contribute the greater proportion of the forage. If allowed to grow too long the stem may become fibrous and past its greatest usefulness, and considerable care is necessary in the spacing of the plants and the time of feeding if the best results are to be obtained. The plants are not as winter-hardy as thousand-head kale, and it is usual to cut the crop before Christmas, although of the two main types cultivated, the purple stem is probably more frost-resistant than the green stem, but the latter gives higher yields than the former under high fertility conditions.

Kohlrabi, which is sometimes not included with the *oleracea* forms but is regarded as a separate species, has a special place among these cruciferous forage crops. It produces a small oval or spherical swollen main stem that sits on the surface of the soil and is essentially of the same nature as the swollen stem of the marrow-stem kale, but its leaf production is very small. The swollen stem, which is sometimes confused with, and likened to, a turnip, may be green or purple according to the variety, and there are early- and late-maturing strains. The growth habit of the plant makes it particularly suitable for folding, while its drought-resisting nature and its high degree of resistance to 'finger-and-toe' disease, which is particularly damaging to many cruciferous crops, give the crop especial value. The crop is grown most extensively in south-east England on medium soils where it is commonly folded, but it is particularly valuable for milking cows and has the advantage that it can be stored in clamps or even allowed to remain in the field for the first part of the winter.

The total acreage of all the *oleracea* forms is not very large, and it is interesting to note that many of them are still regarded as market-garden crops and have not really found a place of permanence or importance in agriculture. It is worth recalling that it was not until the latter part of the seventeenth century and the beginning of the eighteenth century that the oldest of our agricultural cruciferous crops—the turnip and the swede—became widely established in this country. Previous to this time, the turnip and the swede had been regarded primarily as garden crops, and their methods of cultivation had not been regarded as generally applicable to agriculture until the successful rotational farming based on the turnip and red clover in the Low Countries had been introduced and tried in England. The introduction of the turnip, and later the swede, helped to revolutionise British agriculture by stabilising rotational farming and thereby making more intensive crop husbandry possible. This was accompanied by an improved livestock husbandry because of better forage supplies, and while realising and appreciating the part played by the swede and turnip in this great advance, the possibilities offered by the later introduced cruciferous plants need careful consideration.

The turnip, humble in its associations but important in its agricultural traditions, belongs to a separate species of *Brassica* named *rapa*. The crop appears to be of Asiatic origin in its considerable variability, and there are several geographical groups of forms showing many kinds of root shapes and colours. The turnip is characterised by a swelling of the tap root and the short piece of stem below the seed leaves, but there are also forms in the species *rapa* which do not swell in this way but which develop a crown of leaves in the first year. Some of these 'non-bulbing' forms are known as rapes, and are grown for green forage; others are used as potherbs and develop leafy shoots, interesting forms of which are the old 'Thames-side Brassica' and the 'Seven Top Turnip'. Annual forms are grown for seed from which oil is extracted, but these forms are not cultivated in this country as agricultural crops at the present time.

The importance of *rapa* forms in British agriculture is dependent entirely on the turnip, which is the only representative of any significance, although there are two rapes met with in cultivation elsewhere. Although in the early days of turnip cultivation in this country a greater variety of root shapes and colours, including longs and blacks, used to be grown, the number of varieties at present is large but the root types more restricted. The tendency in this country is to concentrate on forms with round, oval or globe-shaped roots with skin colour of the parts above the soil varying between white, green, red or purple. The colour of the flesh may be white, pale yellow or yellow according to the variety, and it is usual to characterise varieties on the three attributes of shape, skin colour and flesh colour. The colour of the flesh particularly, is more than a matter of academic interest because it is associated with the important character of feeding value and also the suitability for different methods of management and utilisation. White turnips include some of the quickest growing, heaviest yielding and poorest feeding value varieties, often having a dry-matter content of only a little over 7 %, and poor keeping quality and frost resistance. These types are the best for infertile soils, and owing to their quick growth they can be sown late or treated as a catch crop, but they must be used directly they are mature. The pale yellows, or soft yellows as they are sometimes called, are also quick growers but have a feeding value superior to that of the whites owing to a dry-matter content of about 1 % higher than the latter. The true yellow turnips are the frost-hardiest, slowest maturing and highest dry-matter types of all the turnips. They approximate to swedes in feeding value and keeping qualities, but the best yellow turnips are not the equal of the best swedes in these respects. Standard varieties in these three groups of turnips are White Globe, Greystones, Red Globe and Green Top White; Fosterton Hybrid, Sheepfold and Centenary; Aberdeen Yellow, The Bruce and Purple Top Yellow.

The common swede varieties cultivated in this country have yellow flesh, but they vary in the shape of the root and the colour of the skin, most being oval to tankard in shape, and purple, bronze or green coloured in the part exposed above the soil. Swedes can be recognised from turnips by their blue-green, smooth foliage which arises from a more pronounced piece of stem (the neck) than is the case with turnips. They are slower maturing, firmer fleshed, of higher dry-matter content and therefore of better feeding value than turnips, but they vary from variety to variety in all these characters as well as in keeping quality, disease resistance, frost resistance and yielding capacity. As a whole swedes do better on soils of higher fertility than do turnips, and they are not a success in dry districts where turnips are used with arable sheep flocks.

The incidence of disease, rate of maturation of the root, the yield of dry matter and the keeping quality are the most important characters of swede varieties and between them decide the value of the crop. To get the best out of the crop the time of lifting and feeding must be adjusted to the time of maturation of the roots, and varieties may be classified as early, midseason or late. Early maturing varieties should be used when they reach maturity because they lose dry weight with keeping; midseason varieties may be kept for some time; while later maturers can be clamped or left in the field, when the winter is not severe, for considerable periods without suffering any material dry-matter loss.

All these characters affect the amount of stock feed available from a crop of swedes and should be considered in choosing the most suitable and economic variety to grow. In addition, there is the matter of the conformation of the root. Some varieties tend to produce an undue proportion of malformed and undesirable root shapes which may seriously reduce the feeding value or the ease with which the crop can be lifted at harvest. The least waste and the maximum amount of good stock feed is obtained when the roots are of good form and generally symmetrical, with freedom from long and multiple necks or fangy branches.

Among the earliest maturing varieties are certain high yielding, but low dry matter purple-top forms with little frost resistance. Best of All, Early Round, Magnificent, Magnum Bonum, Superlative and Majestic are examples of this group. Later maturing, purple-top varieties with a higher dry-matter content are Bangholm, which is also very resistant to 'finger-and-toe' disease, and Aberdeenshire Prize. The green-top varieties are usually the slowest maturing, and have the highest dry-matter percentage, the smallest cropping capacity, the best keeping quality, and are the most frost-hardy. Outstanding of this type is Wilhelmsburger, which shows extremely high resistance to 'finger-and-toe' and to certain other fungal diseases. The only white flesh variety recently introduced into cultivation is Garton's White Fleshed Variety, which is rather low in dry matter but a heavy cropper.

The dry-matter percentage of turnips and swedes varies considerably with the growing conditions including weather, soil and fertiliser treatment, but the variety and the stage of growth and ripeness have the most important effect. Individual roots of any one variety will vary considerably in their amount of dry matter, but the average for the whole crop is the important thing. From the $7\frac{1}{2}$% of the white turnip to the $12\frac{1}{2}$% of the late-maturing swedes, there is a considerable difference but the true value must be reckoned in terms of total dry-matter production which is dependent on the yield of roots as well as the percentage of dry matter. It is generally found that large roots have a low dry-matter percentage, and it is usual for high dry-matter varieties to produce low root yields.

The swede is botanically quite distinct from the turnip, and the confusion by referring to the swede as the 'Swedish turnip', or by alluding to the swede and the turnip as the 'turnip crop' should be avoided. The species name for the swede is *napus*, or *napobrassica*, and it is considered to be of Mediterranean origin and possibly to have arisen by natural hybridisation between the two species *rapa* and *oleracea*. In addition to the swede, the species *napus* includes the swede-like rapes, of which there are two types—giant and dwarf; certain kales like the Hungry Gap; and Colza seed or coleseed which is a type grown for its seed and the oil which it contains. The swede-like rapes are the common rapes of this country and are grown for sheep folding or cutting; they may be taken as a major crop in the rotation or as a catch crop, while the crop can also be used for colza seed, the residue left after expression of the oil being sold as rape cake for stock feed. Rape for forage is very adaptable as regards soil and management, and by suitable adjustment it can be utilised at various times of the year.

There are two true mustards belonging to different species of the genus *Brassica* cultivated in this country. Both are annual plants and each has a special application in the economy of the farm. The black mustard (*B. nigra*), so-called because it has a dark brown seed, is grown entirely for the manufacture of table mustard, which is used as a condiment, but the white mustard (*B. alba*) and the brown or Indian mustard (which is not cultivated in Britain) may be used for a similar purpose. The economic use of the seeds of these brassica species is dependent on the presence of certain sulphur compounds in the seed leaves. When the seeds are ground to a powder and water is added, as in the preparation of table mustard, a compound is produced by the action of an enzyme in the seed which has the characteristic 'bite' and flavour associated with the condiment. The seeds of these plants can also be used for the extraction of oil, the residue being usable as a manure.

Black mustard for seed is grown on a very limited scale, and the whole acreage is confined to certain counties of eastern England. It requires fertile and moist soils but will not stand high rainfall or lack of sunshine if well-ripened seed is to be obtained. White mustard is one of the quickest-maturing crops grown in this country, and can be grown on a wider range of soils than black mustard. It also has more than one use, being grown most extensively as a forage crop or green manure, and therefore is not so restricted with regard to climate. As a forage crop it may be sown alone or mixed with turnips for sheep folding, and it is a common practice to take the crop after a bastard fallow or some early-maturing main crop. It is also used as a green manure for which purpose it is ploughed in before reaching maturity.

The importance of the *Brassica* plants in British agriculture cannot be over-emphasised, and with the Gramineae and Leguminosae contribute the crops that are the basis of rotational farming. The turnip and the swede first impressed themselves on agriculture in this country as cleaning crops for helping to maintain the land from becoming foul with weeds, and as succulent forage crops which contribute stock feed at times of the year when fresh material was in short supply. These two plants still occupy the most important position of any root crop in this respect, and they contribute about three-quarters of the acreage sown to *Brassica* crops in the United Kingdom, and in normal times are approached in acreage only by the potato. Close on a million acres of *Brassica* crops were grown in the United Kingdom, before the 1939–45 war, a certain proportion of which consisted of turnips, swedes and cabbages for human consumption, while there were in addition considerable acreages of these crops grown entirely as vegetables.

The great value of this group of plants from the point of view of their cultivation is that between them they offer such a wide range of different types, one or other of which can be grown anywhere within the range of soils and climates of this country. In addition, the different growth habits make it possible to ensure a supply of succulent forage at practically any time of the year, and the different methods of utilisations, in particular folding, cutting green, clamping or even silage, offer considerable scope for the feeding of all classes of stock. Succulence, bulk, and high digestibility are the chief characteristics of these crops when fed in the proper condition; and although the feeding value varies considerably between the separate crops, and also in the different varieties of some of the crops, they are all characterised by being essentially carbohydrate suppliers. In recent years there has been an increase in the use of turnips and rape as nurse crops for newly sown grass leys and as constituents of 'pioneer leys' and late sown one-year grazing swards.

There is, in truth, a *Brassica* plant available for virtually any kind of growing conditions and for an extremely wide range of farming systems and management. They offer special consideration for the feeding of milk cows, although due care must be taken about the dangers of milk tainting especially with turnips and swedes, and it is safest to feed all brassicas after milking; they have special application in folding and on certain classes of light land are regarded as indispensable for maintaining the soil in a productive condition; they are suitable for all classes of fattening animals and are useful for poultry. Brassicas are, in fact, the crops for intensive farming, for intensive livestock husbandry, and for mixed farming in general. They are normally taken in the rotation where the 'root' break comes in, swedes and turnips for example, usually following and preceding a cereal, as also may all the others when taken as a major crop.

The attention that has been given to the improvement by breeding of the different cultivated forms of brassicas varies considerably. Large numbers of varieties of swedes, turnips, cabbages, brussels sprouts and cauliflowers are in existence, and in many of the purely vegetable types, there are different strains which are in some cases more important in their differences than is the distinction between certain varieties. Some commercially developed vegetable strains, which are the result of many years of painstaking selection by seeds firms, are very distinctive and of especial value, while others are of little economic significance other than being reasonably true to type. As in many of our other cultivated plants, the attention of the breeder has often been concentrated too rigidly on size, whereas in all these succulent vegetables and fodders, feeding value and quality, and in some instances disease resistance and keeping powers, should receive constant attention. There is a vast confusion among the great number of varieties in some of the vegetables and to a lesser extent in swedes and turnips. On the other hand, some of the agricultural crops like the kales and rapes have received little attention from the breeder, and in certain cases the types available are very restricted in character, while there is obvious scope for attempts at improvement.

The cross-pollinating habit of all the *Brassica* crop plants and vegetables, with the exception of the true mustards, makes it necessary to observe certain precautions in breeding and seed growing. Both black and white mustard are self-fertile and are said to be incapable of pollinating one another or any other form of *Brassica*. All the *oleracea* forms intercross more or less readily and produce fertile offspring, so that it is necessary to grow them for seed in isolation. The same may be said of both the *napus* and *rapa* forms, but the practical danger is here not so great because there are not so many undesirable forms of impurity available. Hybridisation between different varieties of swedes or of turnips is, of course, possible, but this is obviously not so objectionable as would be the hybridisation in the *oleracea* group of say cauliflowers and sprouting broccoli, or cabbages and brussels sprouts. The greatest danger in the *napus* and *rapa* forms is the contamination of swedes or turnips with their corresponding rape types. Turnip-like rapes are not commonly grown in this country, and the danger of hybridisation with turnips is small, but there is a real chance of swedes being contaminated with swede-like rape pollen. Such contamination leads to the occurrence of the 'bulbless bolters' in swede crops, while hybridisation with Fill-Gap or Hungry-Gap kales can lead to undesirable mixtures.

Oleracea forms do not hybridise readily with *napus* or *rapa* even when attempts are made artificially to bring about the hybridisation. The possibilities of seed contamination naturally are therefore remote. *Napus* and *rapa* forms interhybridise in all their forms, and the greatest danger is for turnip pollen to contaminate swedes. Such a hybridisation leads to the production of large roots which are usually disfigured by nodules, but the plants are unstable sexually, and because they give little seed, are unlikely to have any lasting effect on a seed stock. But these interpollination possibilities, taking into consideration that not only may members of one species hybridise, but also different species, make it necessary for special consideration to be given to stock maintenance, and degeneration through out-pollination.

Chapter XV

POTATOES

The botanical family Solanaceae, to which the potato belongs, embraces a number of genera which include plants of economic importance because of the possession of a variety of characters. Some, like the potato and the tomato, produce edible parts, although of quite different kinds. Others, like the tobacco and the deadly nightshade ('belladonna') secrete chemical substances (alkaloids) which are used commercially, while several are of importance only because they are wild plants possessed of poisonous properties. Although the family is distributed generally in temperate and tropical regions, it is better represented in the warmer parts of the world than the temperate and colder countries, where particularly in the northern latitudes, including this country, the representation is poor.

Although agriculturally the Solanaceae contributes only a small number of cultivated plants, the two important crop plants —the potato and tobacco—are of great economic importance, although for very different reasons. The tomato is one of the most valuable horticultural crops for outdoor and glasshouse culture in many countries, and has become a widely used fresh vegetable product of special dietetic value in recent years. But in spite of the relatively high cash value of such crops as the tomato and tobacco, the latter particularly being responsible for the expenditure of enormous sums and the source of considerable revenue through taxation, there is no other crop belonging to the Solanaceae which can remotely compare in worth with the potato. This plant has assumed a place of vital significance and importance in the economy of many countries because it produces an edible part, the tuber, which is a staple food for millions of people. It possesses special virtues as a cultivable and edible plant which are possessed by no other food crop, being suitable to cultivation on all scales from the garden to the large mechanised farm, and being characterised by the inestimable virtues of providing a healthy and nutritious staple food, utilisable in a variety of ways at the table, without any more processing than cooking.

The potato is unique among the major food crops of the agriculture of this country in being a native of the New World, and as such has a comparatively short history as a cultivated plant in Britain. All cultivated varieties of the potato belong to the genus *Solanum*, which exists in such a profusion of wild and cultivated species and forms in South and Central America as to leave no doubt that this part of the American continent is its original home. There are so many different tuber-bearing species in this area that it is difficult to say which are wild and which are cultivated, while the existence of closely related wild species which do not bear tubers indicates a great centre of diversity with a long history.

A study of the native civilisations in this centre of occurrence of *Solanum* species shows that the potato has played a vital part not only in feeding the people, but also in the development of religious rites and art. As far as can be ascertained, potatoes were first taken into cultivation in the high inland plateau and areas of the Andes, where the plant provided the staple food of the mountain peoples. Native pottery of the third century A.D. and later, show the potato as the inspiration of ceramic design, and its obvious effect on religion and thought can be followed by the evolution of design and form in the pottery. There is no evidence that the cultivated potato had spread beyond the original centre of Chile, Peru, Ecuador, Colombia and Bolivia before European travellers first reached these countries, in spite of the fact that tuber-bearing forms occur as far north as Guatemala and Mexico.

The botanical and cultural history of the potato in its American home has been the subject of much study in recent years, largely as a result of several plant expeditions which have been organised for the specific purpose of collecting material which might be used in plant breeding. Although it is known that the Spaniards were the first Europeans to become acquainted with the cultivated potato in South America in the sixteenth century, and also that the first potato reached Spain about 1570, it is not clear by what means, or in what circumstances, this introduction to Europe was made. There is no certainty whether this first potato and the one recorded in Vienna in 1588, belonged to the species *tuberosum* which is grown in Chile, or whether it was a different form from the Andes which is usually regarded as a distinct species (*andigenum*). Both types may have been introduced into Europe and subsequently grown there, eventually giving rise to the many cultivated varieties developed on the continent. There is also some uncertainty where the potato described by the English herbalist Gerarde in 1596 came from, although there is little doubt that all the old European and British varieties that were cultivated for hundreds of years were derived from these original introductions.

The great value of the potato as a cultivated plant was not quickly appreciated in Europe or in the British Isles, and it was for a long time regarded more as a botanical novelty. In these islands the Irish were the first to take up its serious cultivation, but it was not until the industrial revolution that any great headway was made in England and Wales, although certain areas in both these countries had developed local potato-growing areas sometime before this. With the passing of the old open-field system of farming, and the increased demand for cheap food from a rapidly growing but underfed and underpaid population, the potato rapidly sprang into prominence as a valuable plant. Not only did it assume considerable importance as a farmer's crop, but it became the great standby for the cottager and the town dweller with any plot of land capable of supporting a cultivated vegetable. The potato showed its extreme adaptability as a crop for the specialist grower, for the mixed farmer, for the market gardener and the small-holder, all of whom cultivated it for the rapidly expanding urban market, while the domestic allotment holder and gardener grew it to supplement the family food supply.

No other food crop in temperate regions has achieved such a position as the potato, because there is nothing to compare with it as an accommodating plant from the point of view of cultivation and convenience for use as an everyday and staple article of diet. It is easy to grow, it is adaptable to a wide range of soils and climates, and above all it is cheap and easy to prepare for the table, providing a nourishing bulk food with a minimum of preparation. There is no expensive processing, it may be eaten in a variety of ways and with a variety of other foods. It is a fresh vegetable as well as a convenient supplier of bulk and energy, and it has a long season of consumption and may be left in the ground or stored for considerable periods of time. Above all, it can produce more food to the acre than any other crop, supplying not only carbohydrate, but also first-class protein, vitamins B and C, and mineral matter. There is little cause for wonder, then, that the potato has achieved such importance, although its very virtues have been the cause of considerable sociological questioning because of the low standards of living and the exploitation of human beings that are associated with it.

The edible portions of the potato plant are the tubers which are produced below ground after the plant has developed its full vegetative green shoots above the soil. The production of subterranean tubers, more than one of which are normally borne on each plant, is encouraged by earthing up the plants in order to bury the basal parts of the stems and stimulate the development of the special stem growths, or stolons, at the ends of which the tubers are borne. Each tuber is the swollen end of a stolon, and is merely a special food-storage stem structure which serves the plant as an organ of propagation and multiplication in place of the seeds of other plants which are normally sexually propagated. The potato plant, which has developed the tuber as a means of reproduction, has largely lost the capacity to produce seeds, and although many tuber-producing kinds also bear flowers, these are in many cases sterile or easily shed.

The potato plant is unique among our major agricultural crops in showing this extreme specialisation in vegetative multiplication and food storage. The formation of tubers is dependent on there being a surplus of carbohydrates over and above the amount necessary for growth, and these carbohydrates, the most important of which is starch, are transported below ground where they accumulate in the tubers. As a result of this local concentration and accumulation of reserve food, the tuber is stimulated to grow and increase considerably in diameter. The whole of the internal tissues become modified for storage, except the small amount of tissue necessary for transporting food materials, and the thin skin on the outside which has a protective function and in which are situated the buds ('eyes') which give rise to the new shoot and root system. It is this combination of local concentration of food reserves combined with the edible nature of the tissues after the simplest of preparation by way of cooking that has made the potato the most important world crop in terms of bulk food yield for human consumption.

Although potatoes are now cultivated in most countries with temperate climates, there is little international trade in this crop because many countries produce all they require by their own exertions, and the potato is an expensive, bulky and difficult product to handle in large quantities and over any great distances. These handicaps to international trade are due to the nature of the potato tuber as a commercial product; it is a living and succulent vegetative part of the plant which is susceptible to many forms of damage and disease and consequently is the object of many trading restrictions and regulations. It may possess between three-quarters and four-fifths of its weight as water and is consequently expensive to handle in relation to its food value. A certain amount of international trade exists where some countries, such as Great Britain, are prepared to pay relatively high prices for the crop when it is out of season at home. But for the most part the potato is home-grown to provide a regular part of the human diet, and as such it has acquired a position which rivals that of wheat in certain countries. Indeed, reckoned in terms of total production in tons of produce, potatoes head the list as a world crop, and its uses as a stock food and a raw material for the commercial manufacture of starch and alcohol must also be considered when assessing the value and uses to which it may be put. In Great Britain the great value of the potato crop is as a human food, and the high cash value of the home-grown crop is due to this market.

Some potatoes are grown on most farms in this country, even if only for home use and for feeding stock. But commercially produced potatoes for the market, on which the cash value of the crop depends, although distributed throughout nearly all the important agricultural tillage areas, are grown most extensively in certain districts and counties. The big potato-growing counties are Lincolnshire, Yorkshire, Cambridgeshire and Lancashire, with highly intensive and specialised areas in the coastal districts of Wigtownshire, Ayrshire and Lincolnshire. Thus, although potatoes may be grown successfully over practically the whole country, the bulk of the country's output is localised, though scattered. Certain areas have small but commercially valuable potato-growing districts, such as the early potato growing of south-west England, where the climate and the soil are especially suited to this type of production.

Potatoes may be grown successfully on practically all types of soil, but they are not suited to the heaviest and wettest soils, nor do they do well on light, dry soils of low fertility which are liable to be affected by drought. Good fertility, a free working soil and a plentiful water supply are necessary for good results, and the heaviest crops are only obtained when there is a liberal humus content, either naturally as in fen soils, or by the application of dung or green manure or the ploughing in of turf. The place of potatoes in the rotation is decided primarily by these needs, but also by the fact that the crop is a good cleaning crop and an excellent preparation for other crops which require thorough soil preparation. In practice, potatoes are commonly taken after grass, after a grain crop which is usually oats, after a previous potato crop, or after a red clover or other green manure crop. In some specialised potato-growing areas, particularly where early varieties are the speciality, unusual rotations are practised which are designed often to fit in with intensive vegetable production.

Potatoes are unique in our farm crops in that the so-called 'seed' is not a sexually produced true seed, but is a vegetative off-shoot of a bulky and more perishable nature than a seed. This off-shoot, which is simply a small-size tuber, is the sole means of propagation for agricultural purposes, and its character and peculiarities are responsible for many special features, some advantageous and some disadvantageous, in the handling of the crop. The first and most far-reaching of these is the disease question, and so-called 'running out' or degeneration. It was thought for many years that the continued vegetative propagation of potato varieties was in itself responsible for the decline in yield and vigour which was obvious in some varieties, and that sexual reproduction would be necessary to restore them to their original vigorous condition. It is now known that this is not the true explanation, but that progressive infestation with certain viruses is the cause. Once a plant becomes infected with a virus, provided the plant is not killed, the virus permeates every part, so that although infection takes place through the leaves, the tubers become infected. When tubers for seed are taken from such a plant it means that the grower is starting with a virus-infected plant, and the vigour is reduced. If this practice is persisted in season after season, the crop becomes progressively more and more infected and the yield will continue to drop until it reaches uneconomic levels.

This then is the principal disadvantage of tuber propagation. The only ways of avoiding the trouble are either to grow stocks of seed tubers which are virus-free, or to grow varieties which show some form of resistance or immunity. The first method is the one at present commonly followed, and stocks are grown for seed in areas where virus infection of certain types is at a minimum, and certified stocks are supplied to growers after they have been inspected and shown to reach the necessary standard of freedom from viruses. Once a certified stock has been distributed to growers it will maintain its vigour only in so far as it remains free from virus infection, and whether or not this will happen will depend on the part of the country where the stock is grown. It is possible to maintain stocks in a healthy condition for a much longer period in the north and west than in the south-eastern parts of the country, and seed should be changed more often where the chances and risk of virus infection are greater.

But virus diseases are not the only infections of seed tubers which are responsible for loss of crop due to planting diseased tubers. A gappy plant stand in the field may result from a virus, or from fungal or bacterial infections such as Skin Spot, Dry Rot, Black-Leg and Gangrene. Some diseases may not only cause failure or early death of the plant, but as in the case of Blight,

PLATES
28–36

TESTING THE GERMINATION OF GRASS SEED

PLATE 28

The actual testing of the germination capacity is done by counting the number of seeds which germinate in a prescribed time under definite moisture and temperature conditions. Although the behaviour of the seed under such laboratory tests is not always a true guide to the performance in the field, seed samples that do not come up to a satisfactory standard can be discarded as unfit for commercial purposes.

CONTROLLED BEE POLLINATION OF SAINFOIN

PLATE 29

Controlled pollination of insect-pollinated species such as lucerne and sainfoin is obtained by using movable cages which prevent unwanted pollinating insects visiting the flowers. Lucerne and sainfoin are pollinated by bees, and plants may be self-pollinated by isolation in a cage, or they may be hybridised by enclosing two or more plants as required. Bees free from pollen are introduced into the cages when flowering starts.

PLATE 30

Many domestic varieties of potatoes are sexually sterile in one way or another, and breeding work is considerably handicapped by the difficulties involved in obtaining hybrids. Inter-varietal hybridisation has, however, resulted in the production of many new varieties in the potato, as in other crops, and sexually fertile seedlings are being used for breeding purposes. Hybrids between domestic varieties and allied species of *Solanum* are also being made to try and solve some of the problems of potato improvement, particularly that of resistance to Late Blight. Large numbers of hybrid seedlings are raised in glasshouses for testing resistance to Blight, and the stocks are kept free from virus infection by making the glasshouses insect proof.

HYBRIDISATION OF WHEAT

PLATE 31

Controlled pollination, involving either self- or cross-fertilisation is an important method in plant breeding and in the maintenance of seed stocks. In the normally self-pollinated cereals, the flowers are first emasculated if it is intended to hybridise. After emasculation, the ears are enclosed in cellophane or parchment bags before and after transfering pollen from the male parent. The grain resulting from such a cross-fertilisation is thus of hybrid origin, and is used to grow the first generation hybrid, the progeny of which is the second generation hybrid. It is from the second and subsequent generations of hybrids that selections are made of new and improved types.

PLATE 32

The breeding and agricultural value of individual grass plants is first judged by dividing the selected plants into separate tillers and planting each. These give rise to a series of vegetative progenies, the whole vegetative family being called a 'clone'. The true characters of the originally selected plants may be studied and judged more satisfactorily by examining the characteristics of the clone than from the single initial plant. In this way material for the building up of new strains of grasses can be provisionally selected, but the breeding behaviour has to be tested subsequently from sexually produced progenies.

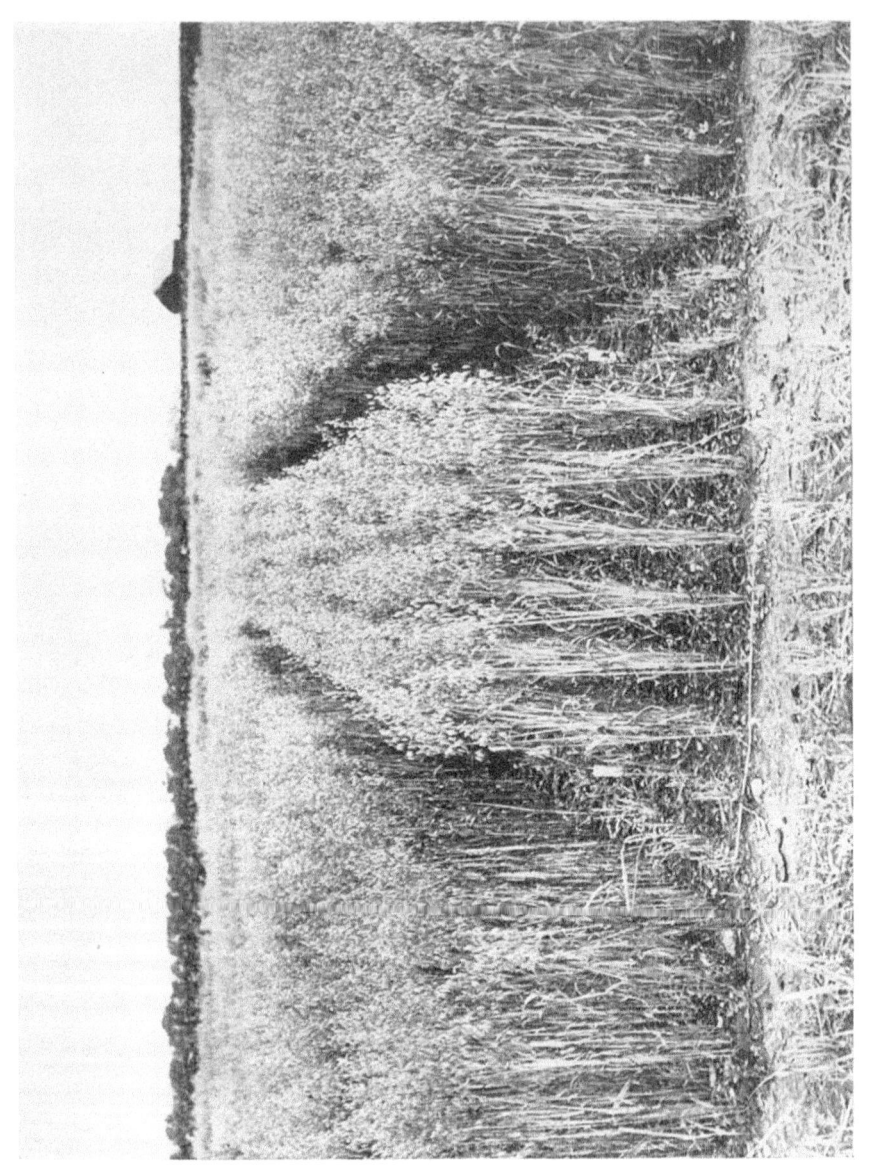

FIELD-TESTING NEW OAT HYBRIDS

PLATE 33

Selected cultures from hybrid progenies in such self-pollinated crops as oats, wheat and barley are re-selected from year to year until they are apparently true breeding for all observable characters. The chosen cultures, which possess the desired combination of characters, are then tested in small-scale yield trials where they can also be observed still more critically for all important characters. Each culture is replicated several times in the trial which may consist of a large number of uniform-sized plots, depending in number on the number of cultures and the number of replications of each. At harvest the border rows between each plot are discarded, and the remainder cut out by hand, the produce from each plot being kept separate. In this way the yielding capacity of the cultures can be calculated with considerable precision under the particular conditions of soil and climate which characterise the trial.

CHECKING THE IDENTITY OF BARLEY FROM TRIAL PLOTS

PLATE 34

The breeding of improved varieties and the maintenance of existing stocks requires careful observation, recording and testing. The produce of isolated small plots needs to be accurately labelled and checked at harvest, and only sheaves of the same stock should be allowed to come into contact with one another. Material from different cultures, selections and varieties are kept separate to avoid contamination so that pure stocks can be maintained for multiplication or field trials. These precautions only prevent mechanical mixing in the field, and they must be followed by similar care in threshing when there is considerable danger of contamination. But the greatest care in management and handling cannot prevent stocks from becoming impure through natural out-pollination in the field, and if these are to be prevented special isolation must be provided.

MULTIPLICATION OF A NEW WHEAT STOCK

PLATE 35

Initial field multiplications of new stocks are inspected carefully and may require special attention to prevent bird damage. Such stocks may be the result of 10–15 years' work on the part of the plant breeder, and care is needed to avoid the waste of years of work. New stocks which have passed the trial stage and which are to be put on the market as new varieties, as well as certified stocks of old varieties which are being multiplied, are all subjected to this type of management and inspection before being released to the seed trade or the grower.

OBSERVATION PLOTS OF HERBAGE PLANT STRAINS

PLATE 36

Strains of herbage plants appear on the market often with no distinctive name other than the country of origin. In some species there is considerable confusion owing to the large number of strains, and even the nationality of the strain is not sufficient to characterise its behaviour. The most elementary comparison of strain characters can only be obtained after years of observation, and the comparative behaviour of strains from different sources when grown together under similar conditions is the only reliable means of obtaining the first judgement of their agricultural value.

when infection takes place in the growing crop, the yield is reduced, the tubers are blemished, and disease is set up in the clamp which may cause considerable loss. The susceptibility of the tubers to diseases, the dangers of frost and mechanical damage which cause loss of seed tubers and marketable produce in storage and in clamps, and the bulkiness and general inconvenience of handling all classes of tubers are then the great disadvantages of vegetative propagation in the potato crop.

The only advantages to set against these serious troubles are that, provided the stock can be kept healthy, it is possible for a grower to keep this stock indefinitely without much trouble from it becoming mixed or from throwing 'off types'. In hybrid varieties of crop plants which are propagated sexually by seed, there is always a greater chance of 'off types' appearing, and if the crop is a cross-pollinated one, there are added and greater difficulties in maintaining a stock reasonably pure. Potatoes are free from these troubles, but they are subject to other sources of variation, such as the tuber 'sports' which arise occasionally, and also undesirable types known as wildings and bolters which appear more in some varieties than others. It may be noted in passing that there are certain advantages to the potato breeder in this vegetative propagation, but to off-set this, the loss of sexual fertility in the potato can be a source of serious embarrassment to the breeder engaged in trying to improve this crop.

The growing and the commercial value of the potato crop are consequently bound up intimately with the tuber-bearing characters of this plant. The gradual selection and development of new varieties in cultivation have concentrated on tuber characters, and it seems that in the process the potato has lost the capacity for sexual reproduction to a greater or less degree according to the variety. The tuber characters and peculiarities are of the utmost importance to the grower from the point of view of the handling of the 'seed', and special methods of seed growing and preparation for planting have had to be developed. Healthy seed from good stocks is as important as the choice of the right variety, while the careful preparation by sprouting of the tubers for planting is an important means not only of

145

ensuring that only vigorous tubers are used, but also of obtaining a heavier and earlier crop.

But the tuber is also the marketable product and its characters are the final determiners of the uses to which the crop can be put. The potato is essentially a bulk carbohydrate food, which if eaten in sufficient quantities can supply all the energy, protein and vitamins B and C requirements of the adult human being. Potatoes are normally not used in this way, however, and being deficient in protein for young people, and also being inadequate in most minerals and other important vitamins, they cannot be regarded as a complete food in themselves. Varieties, however, vary in their nutritive value, some having a higher dry matter, a higher protein, or a higher vitamin C content than others, although the time of lifting, the weather during the growing season, the soil type and manuring will all have their effects on the amount of dry matter and the proportion of protein in the tubers.

Apart from the differences in the composition of the tubers, potato varieties show characteristic differences in certain tuber characters which have a most important effect on the market value and local popularity of varieties. Some of these characters are entirely superficial in that they do not affect the intrinsic worth of a variety, while others are more fundamental because they influence the efficiency of usage. It does not matter intrinsically, for example, whether a tuber is round, oval, long or cylindrical; nor is it of any material value whether the skin is coloured or white, rough or smooth. These characters do not affect the feeding value or the cooking quality, but like the colour of the flesh they are important because there is a definite public preference, which varies in different parts of the country, for varieties possessing particular tuber characters.

Far more important characters are those which affect the ease of handling in the kitchen and the cooking quality. Varieties with uneven-shaped tubers, and with deep eyes, are subject to more waste in preparation for the table than those of even shape, freedom from lumps and hollows, and with shallow eyes. On cooking, the texture of the flesh varies between varieties although the method of cooking can affect this character. The popular taste is for a floury potato which keeps a good white colour on cooking, and for some circumstances it is obviously desirable to have a potato that can be reheated without blackening. It is generally agreed that the type of soil affects the cooking quality, and the highest priced potatoes are grown on limestone and red soils, while the poorest are obtained from peaty soils. Manures and fertilisers affect quality when deficiencies or over-dressing adversely affect the balance of growth, and generally speaking the quality improves as the crop matures.

All these characters are principally of importance in the main-crop varieties which are allowed to grow to maturity. Early varieties which are lifted immature are not subject to the same judgement because in their case these characters do not have the same chance to show themselves. The important thing about an early variety is that it should produce as heavy a marketable crop as early in the season as possible. Out-of-season and very early potatoes are virtually a luxury product, as far as price is concerned, and are treated as such by grower and consumer. This is in sharp contrast to the maincrop of potatoes which becomes the cheapest food on the market in this country, and may in seasons of plenty be not worth marketing for human consumption and is used for feeding stock when the market is glutted.

Tuber characters affect the grower only in so far as they are responsible for influencing the demand on the market for a particular type. As far as growing the crop is concerned, it does not make any material difference whether a variety has deep or shallow eyes, or a coloured or white skin. The growers concern is to produce a profitable crop, and this is dependent on his growing high yields of marketable tubers for the domestic consumer. Yield is a varietal character, as also is the proportion of tubers of ware size, which is the size necessary for the domestic market. Some varieties produce large tubers, others comparatively small ones, but both the absolute size and the relative proportions of large and small tubers will depend very much on the soil fertility, the manuring, the size of the tuber that is planted, and the spacing of the plants. A grower who is concentrating on grow-ing ware-size tubers, plants seed tubers of about hen's egg size, while a grower whose principal interest is to supply seed-size tubers for selling to commercial growers, plants ware-size tubers and spaces them closer in the drills. Within limits the greater the weight of tubers planted, the heavier will be the yield, but the proportion of different size tubers obtained is important and it is not profitable to increase the yield simply by adding a large amount of small tubers ('chats') which are not marketable.

Although the choice of the appropriate variety is the first concern of the potato grower in planning the cultivation of profitable crops, it is scarcely less important that he should obtain healthy and reliable stocks. In no other crop is the question of health so important with regard to the 'seed' which is to be planted. The reason for this has been explained as due to the susceptibility of the tuber to diseases, and the devastating effect of these on the yielding potentialities of the crop. The most important of these diseases are the viruses, and it is largely because of these that the elaborate schemes for growing and certifying potato stocks for seed have been developed. In the different countries of the British Isles various forms of certificates are issued, but all are based on the important considerations of as low a percentage of virus infection as possible, a high degree of purity, and freedom from undesirable forms such as wildings and bolters. These schemes are of the utmost importance in maintaining the standards of the stocks and the yields of potatoes in this country. Every grower should realise the importance of the schemes and the significance of the different kinds of certificates, because in no other crop is there such a wastage every year through growers continuing to use disease-ridden stocks of poor yielding capacity.

The whole problem of disease is consequently of the greatest importance in potato growing. The response of all the important varieties to the various diseases which reduce yields or blemish tubers are now known. For example, all varieties show susceptibility to Late Blight, although some are more susceptible than others; some varieties are immune, while others are susceptible to Wart; there are varying degrees of susceptibility to Dry Rot; and the response of varieties to the more serious virus diseases varies considerably. Therefore, apart from the fundamental question of obtaining healthy and sound stocks, the grower needs to consider the question of variety as a means of combating certain diseases. Where neither of these methods proves effective there is the use of chemical fungicides for such diseases as Dry Rot and Late Blight, while field cultivation can also play its part for certain diseases. Good keeping quality, which is so important in varieties which are stored in clamps for any length of time, is largely bound up with the resistance of the variety to various tuber infections.

The number of potato varieties that has been cultivated in this country during the last 100 years is very great indeed. Most of these varieties have been short-lived, largely because they became infected with virus diseases and 'degenerated', and their places were taken by new varieties. There has always been something of a 'fancy' in potato varieties, also, and many new varieties have appeared for a short time because of some more or less attractive peculiarity, and then have passed out of cultivation. Potatoes have further been subject to the problem of 'synonyms' more than any other crop, and there has been considerable confusion with regard to the identity of some varieties because some of the well-known and popular varieties have appeared on the market under many different names.

Apart from yield, cooking and keeping quality, and disease relationships, the important grower's character of potato varieties is the time of maturity. There is a wide range of length of growing season in the many varieties in cultivation and it is customary to group varieties into classes such as first earlies, second earlies, early maincrops and late maincrops. The number of classes can obviously vary according to the standards chosen and the range of varieties available, but owing to the importance of this character with regard to the growing and marketing of the crop, it is at least necessary to distinguish between earlies and maincrops. It is usual, also, always to refer to the fact whether a variety is immune or non-immune to Wart disease, because of the restrictions that have been in force with regard to this important disease.

Commercial potato growing in the British Isles is mainly concerned with the production of maincrop varieties for human consumption, and the country is self-supporting with regard to this commodity. In England and Wales the most commonly grown varieties are Arran Banner, Majestic, King Edward VII and Doon Star, while in Scotland, Kerr's Pink, Golden Wonder, Redskin and Gladstone are the most popular. But each country grows all these, and many more, varieties, although some make only a small contribution to the total production.

Arran Banner, Majestic and Kerr's Pink are very heavy cropping varieties when grown under good conditions, Majestic being the earliest maturing of the three and having long oval tubers in contrast to the round shape of the other two. Kerr's Pink has a pink skin and is more popular in the northern part of the country than in the south, where the white skin varieties are preferred, in spite of the fact that Kerr's Pink is of better cooking quality. Both Arran Banner and Majestic produce very large tubers on fertile soils and they cannot be regarded as first-class culinary varieties, particularly as there may be considerable waste with Arran Banner because of its uneven shape, and Majestic is very prone to blackening when reheated, or even on first cooking. In spite of its defects in cooking, Majestic has been a prominent maincrop variety in this country since about 1911, being followed some four years later by Kerr's Pink, while Arran Banner was first marketed in 1927.

Of the good cooking quality varieties, King Edward is outstanding because it has maintained its popularity since the beginning of the present century although it is not immune to Wart and is very susceptible to Late Blight. King Edward owes its position to the excellent quality and good tuber characters which are so characteristic and easily recognisable. The variety is not a heavy cropper, but the tubers have good keeping characters and they cook exceptionally well. The oval tubers, white-skinned except for the splashed pink colour, are liable to be confused with those of the more recently introduced variety Gladstone, but the latter has a tendency to be rough-skinned and to develop hollowness in the tuber, as well as being a higher yielder. The oldest of the good cooking quality varieties is Golden Wonder, which has survived since the latter part of the nineteenth century. This variety has kidney to pear-shaped tubers with a russeted skin and white to pale yellow flesh and possesses the valuable character of Blight-resistance in the tubers. The tuber feature of white to pale lemon flesh is possessed by Doon Star, a very recently introduced variety of good cooking quality and characteristic purple splashed skin around the eyes. This variety crops well, but has the defect of being a poor keeper.

Arran Pilot is the most extensively grown early variety in England, and Epicure the most popular of this group in Scotland. Epicure has been in cultivation since the last decade of the nineteenth century and owes its reputation to its capacity to give comparatively high yields early in the season and its greater resistance to frost than any other variety in cultivation in this country. Its tubers are rather irregular and indented, it has a white skin which turns a pinkish colour, and although a high yielder it is only of fair cooking quality. Arran Pilot has only been a popular early variety since the early 1930's; it has long oval-shaped tubers with a white flesh and white skin which shows some bluish colour around the shallow eyes. Arran Pilot yields well, but the cooking quality can only be regarded as fairly good.

The other popular but less widely grown early varieties are Eclipse, Ninetyfold, Sharpe's Express and Duke of York, all of which are non-immune to Wart disease and of good cropping capacity, although Sharpe's Express is inferior to the others in this respect. These varieties also have oval or long tubers, sometimes pointed or pear-shaped as in Sharpe's Express, but they vary considerably in cooking quality and flesh characters. Duke of York, which has a yellow skin and flesh is the best quality of the four, with Sharpe's Express, also a yellow-fleshed type, next. Ninety Fold, which is very susceptible to Late Blight and Dry Rot, is a poor cooker, as also is Eclipse.

Although new potato varieties sometimes arise by 'sports' or 'mutations' by far the greatest number of varieties at present in cultivation have been produced as a result of hybridisation. One of the great handicaps for the plant breeder working with potatoes is that so few varieties produce good pollen, while in many cases it is difficult to keep the berries on the plant until they are ripe. Hybridisation between varieties to combine desirable characters has been the most successful method of improvement, but there are considerable limitations to the improvement which can be obtained in this way. In addition to the limitations set by sexual sterility, there is also the fact to be faced that many valuable characters appear to have been lost in the development of our domestic varieties.

With the discovery of the many new species and forms in South and Central America, new opportunities for improvement have presented themselves. Some of these species have been known for many years, and it was found that certain of them possess such valuable characters as resistance to Late Blight and frost. The task of transferring resistance to Late Blight from otherwise worthless species to the domestic potato has been the object of attention for some years and considerable progress has been made. There is further hope that by using appropriate species other valuable characters including higher protein and vitamin content, may be incorporated in domestic varieties, and by this means the economic and agricultural value of the potato crop be greatly increased.

Whatever may be the future developments in the production of improved kinds of potatoes, and regardless of any major changes in the position of other tillage crops in the agriculture of this country, the potato appears to have acquired a permanent place in the food production of the British people. The recorded acreage in England, Wales and Scotland prior to the 1939–45 war was approximately half a million, and during the war this area was doubled. Potatoes are the only staple vegetable food product in which this country can be easily self-supporting without upsetting the balance and economy of the agriculture, while its appeal as a prominent feature of the vegetable garden and allotment adds considerably not only to the area devoted to its cultivation, but also to its unchallengeable position as a human food.

Chapter XVI

SUGAR BEET AND MANGOLDS

Although sugar beet and mangolds occupy such different positions in the economy of British agriculture, and of the individual farm, their close botanical relationship and characters brings them together when considering the cultivated plants of the farm. Not only do these two crops belong to the same botanical family—the Chenopodiaceae—but they are also usually considered as comprising one species of the genus *Beta* which is the only important genus, agriculturally speaking, of the seventy or so genera in the family. The Chenopodiaceae is a family of plants of world-wide distribution and is represented in the British Isles by six genera of herbaceous or low shrubby native plants which are usually found in coastal districts and maritime areas. Some are particularly associated with salty soil conditions and grow on or near seashores, on cliffs, sand belts and estuarine marshes, while others are found inland as common weeds of cultivated land or as inhabitants of waste places. The common oraches and the goosefoot, or fat hen, are the most widely known native plants of the family in this country, while the plant known as all-good, or good-king-Henry, is an example of a wild plant of our countryside which was once a commonly cultivated potherb in many rural districts. Some of the plants in other genera of this family are used in a similar way, the true spinach being an example of an introduced plant to this country which is usually cultivated as a garden vegetable.

The genus *Beta* includes something like ten species, but only one of these is of agricultural interest because it includes the mangold and sugar beet, although other useful cultivated plants such as the red table beet, the perpetual or spinach-beet, the seakale-beet and the chards also belong here. This species, named *vulgaris*, is very variable and exists in a great number of forms in various parts of the temperate regions of the northern hemisphere. Around the sea coasts of this country and of continental Europe, Asia, North Africa and India there occurs the plant known as the wild sea beet which is considered to be a subspecies (*maritima*) of *vulgaris*. This sea beet is very interesting because when it is grown as a cultivated plant it may be found that seeds gathered from wild plants produce many kinds of plants which resemble sugar beet, mangolds, table beet and the leafy forms such as the chards and the spinach-beets. It is for this reason that it is commonly asserted that the sea beet is the wild plant from which these cultivated plants have originated, but as all the cultivated plants and the sea beet readily cross-fertilise, and the pollen is carried by the wind over considerable distances, there is always the possibility that colonies of wild beet growing near cultivated land have been cross-fertilised with a cultivated type of some form or another.

The oldest cultivated plants of *Beta vulgaris* appear to have been the leafy spinach and chard forms and the red table beet, which were grown as vegetables in countries bordering the eastern Mediterranean and in Asia Minor before the Christian era. The mangold seems to have been a more recent development and may have arisen from natural hybridisation between the table beet and the leaf type after these plants had been in cultivation for a considerable period. The development of the sugar beet as an agricultural plant has taken place almost entirely during the last 200 years, the possibilities of using this plant for the extraction of sugar having first been shown in Germany in 1747. It is not known with any certainty what the origin of these original beets was, but they were a mixed lot of red and white types more akin to mangolds than the sugar beet at present in cultivation. Some years after the original work on sugar extraction, a type known as the 'White Silesian beet' was developed. This beet resembled the modern type and was for all practical purposes the first true sugar beet and the mother stock from which all the European stocks appear to have been developed.

The interesting feature of these cultivated plants is that they all appear to have arisen from the very variable *Beta vulgaris* whose original home is probably western Asia and Asia Minor. The original wild types taken into cultivation were the leaf forms, as vegetable and medicinal plants, but the extraordinary range of vegetative form in this species later gave rise to the introduction of the root types into cultivation, with the development of the red table beet, the mangold, and the sugar beet. These types differ essentially from the leaf types in that instead of having a substantial and usually fanged or branched tap root, they show an abnormal swelling of the main root, which does not develop strong lateral branches, and may in some forms grow partly out of the soil, by reason of the enlargement of the stem below the seed leaves as is the case in the swede. There is an almost endless variety of these swollen roots, and, with the less variable leaf types, it is one of the most interesting examples in our farm crops of the possibilities of selecting a wide range of cultivated plants from the variability offered by a single botanical group of plants usually regarded as one species. The *Beta* group, therefore, resembles certain of the *Brassica* crops particularly in the development of edible vegetative parts which are essentially storage organs of root origin.

The mangold and sugar beet, which are the only plants with which we are immediately concerned as agricultural crops, are biennial plants and show certain peculiarities which give them their economic value. When sown in the spring the young plant, if given sufficient space, increases the girth of its root by the abnormal activity of its tissues, which by the successive development of new cells capable of dividing rapidly, forms a series of concentric rings of alternating conductive or vascular tissue and storage tissue. The large crown of leaves is very active in sugar formation and during the first season of growth this sugar is transported to the root where it is stored in the outer part of each ring of conductive tissue, and to a lesser extent in the cells between each ring. When the root has reached its full size towards the end of the summer, sugar continues to pass into the root, so that in the case of the sugar beet particularly, an excessive concentration of sugar is stored in the root tissues. This capacity to store sugar makes sugar beet the only crop of temperate regions which can be grown commercially on the farm for the industrial production of sugar, and with the tropical sugar cane supplies the overwhelming bulk of the sugar consumed in the world. The mangold owes its importance to the large, succulent and nutritious root which is developed and used entirely as a stock feed, for which purpose it is peculiarly suited for use late in the winter owing to its adaptability for storage in clamps after lifting. In this respect it fulfils a function that no other succulent fodder can replace.

Both sugar beet and mangolds are suitable crops to grow over a wide area of arable farming conditions in this country, but the greatest acreages are in the southern districts of England because both require warmth and sunshine for their most successful development. Neither crop is commercially successful on extremely heavy or very light soils, in the former case because of difficulties in seed-bed preparation, in the latter case because of drought. Easy working and well-cultivated soils with retentive subsoils are the most suitable, a good depth of soil being important for adequate root development and avoidance of malformed and fangy roots, a defect which adds to the difficulty of handling during lifting in both crops, and lowers the commercial value of the roots in the case of sugar beet. Mangolds occupy varying proportions of the root break according to the suitability of the conditions to their cultivation and the type of stock kept on the farm. Sugar beet may similarly form part of the root acreage, or in the intensive beet growing areas may constitute the total roots grown.

The position of mangolds in British agriculture has undergone many changes since its introduction from Germany towards the end of the eighteenth century. In 1911 over 500,000 acres were cultivated in England and Wales, but by 1939 this had shrunk to 309,000 acres of which 97% was in England, and chiefly situated from the midlands southwards. The great value of the crop is that it can produce a greater amount of food for livestock rations in the late winter than any other root crop; it may also be stored into the early summer and is a safer and more reliable crop than swedes or turnips in areas to which it is suited. The feeding value is high if the right strain or variety is chosen, and the roots can be fed to all classes of stock, being particularly valuable for cattle.

There are many varieties of mangolds, but these may be grouped into the four main types called globes, tankards, intermediates and longs according to the shape of the 'root', which has a higher proportion of stem in its structure than does sugar beet, and has a greater amount of 'root' growth above the soil. In addition to the shape of the storage organ, varieties differ in the colour of the skin, which may be white, yellow, orange or red, and in the flesh colour which may be of similar colours and developed to varying degrees and intensities in different parts of the tissues. The feeding value of the root depends on the percentage of dry matter, and, contrary to common belief, there is no association between colour and the amount of dry matter. The only easily visible character of the plant associated with dry-matter content is the size of the 'top', or foliage crown; those varieties with a small top having the lower dry matter.

The principal characters which distinguish varieties are, then, the shape, size, colour and dry-matter content of the root, and the size of the tops. From the grower's point of view the important problem is to produce as great a weight of food to the acre as cheaply as possible, which means, in effect to grow the variety which produces the greatest amount of dry matter with the least difficulty in handling. The highest yields of dry matter are not necessarily given by the varieties with the heaviest yields of roots, and it is obviously not as economic to handle a high yielding, low dry-matter variety, as a lower yielding, high dry-matter variety. The great importance of this may be seen when it is realised that some of the low dry-matter varieties have less than three-quarters the dry matter of the better ones, which in spite of lower yields of roots may actually give considerably higher yields of dry matter. It is also important to have a convenient root shape for easy lifting, and the modern tendency is to grow varieties with well-shaped, unfanged roots that have as high a proportion as possible above the surface of the soil. Although these root characters are the primary consideration in choosing a variety an important consideration may also be the size of the tops. Although this character is associated with the dry-matter content, it may also be decisive under certain growing conditions, as for example, the greater suitability of large tops where frosts are liable to be severe, and of small tops on fen soils where too much top is liable to be an embarrassment.

With the great number of named varieties of mangold of doubtful significance in their agricultural differences, little more can be done here than to draw attention to some of the important type characters, although certain varieties show preferences for particular soils and growing conditions. The long types, which are very heavy root yielders under suitable conditions, require the deepest soils and do well on fens and peat soils, whereas the white-fleshed globes are vigorous growers and produce heavy yields of rather low dry matter on the poorer soils. For general purposes, yellow-fleshed globes and red intermediates are usually very satisfactory types, while the white-fleshed tankard variety Kirsche's Ideal has given outstandingly good results in recent trials. Where tankard varieties are preferred, Kirsche's Ideal should be the choice, and the golden tankard types should be replaced by this variety because of their low dry-matter yields, in spite of their relatively high dry-matter content. The grower should choose varieties from among the high dry-matter content varieties, and by experiment select the highest-yielding variety under his particular conditions.

Although there is reference to a 'white beet' in Roman times, this plant was used as a vegetable and animal fodder, and it was undoubtedly quite different from the white beet cultivated for sugar extraction at the present time. The development of commercial sugar production from the beet dates really from the Napoleonic Wars, when the crop was encouraged and subsidised in France, and it was European initiative and enterprise from the early nineteenth century that fostered the industry. Practically every country on the continent has spent large sums of money in establishing sugar beet as a farm crop and in building up a beet-sugar industry by providing the necessary factories. Although the crop was introduced into this country in 1832, and several subsequent attempts were made to popularise it, there was no modern factory until 1912. To-day, after many vicissitudes and changes in fortune, the beet-sugar industry comprises the cultivation of over 400,000 acres of sugar beet in Great Britain and the utilisation of eighteen factories to produce about half a million tons of sugar.

Sugar beet has now assumed the proportions of a world crop of great economic importance, sharing with sugar cane the overwhelming proportion of the total sugar production. The great increase in sugar consumption in the world during the last fifty years (not counting the temporary set-backs of war) has been made possible by the extended and more efficient cultivation of these two sugar-producing plants, and the tendency has been for a gradually increasing proportion of the world's crop of sugar to be produced by sugar beet. In the 1850's, over 90 % of $2\frac{1}{2}$ million tons of sugar was the product of sugar cane, while in 1939 two-thirds of the 29 million tons was due to sugar-cane cultivation, and one-third to sugar beet. This vast production of sugar has become in recent years one of the greatest agricultural undertakings in the world, and the great increase in beet-sugar production has had important repercussions on the agriculture of many temperate countries, particularly that of continental Europe and Russia. But sugar-beet cultivation has had considerably wider effects on the economy of nations because of the complications of government subsidies, import duties and excise taxations which have placed sugar production in a unique position as an agricultural industry owing to its political and international interests.

These matters are important when considering sugar beet as a crop in British agriculture, because this country is one of the greatest consumers and importers of sugar in the world, and beet-sugar production cannot compare in cheapness with cane-sugar production. The total production of beet sugar in this country cannot hope to approach the consumers' demands with the industry at its present size, but the crop is one of the few cash crops grown on contract and is a valuable component of the economic structure of our agriculture. In this respect it is important to add that although sugar beet can only survive as a producer of sugar, it has other considerations to its advantage such as its value as a cleaning crop, the feeding or manurial value of the 'tops' and beet pulp, the commercial value of the molasses which are a by-product of the factory and the employment that is directly and indirectly given by the operation of the factories. Sugar beet cannot be regarded in the economy of the farm merely as just another root crop, therefore, but because of its peculiar position in relation to contract growing, guaranteed prices and close association with industrial processing it has advantages and relationships which are unique.

Sugar-beet growing is essentially part of the series of industrial processes which culminate with the refined sugar in the bag. The efficiency and prosperity of the beet-sugar industry are dependent in the first place on the success of the grower in producing as great a weight of easily extractable sugar as cheaply as possible. The grower is paid for his beet on the basis of root tonnage and sugar percentage, and the factories must be assured of a steady supply of roots in good condition which are capable of yielding the maximum amount of sugar with the least possible weight of roots handled. In the case of sugar beet, therefore, the production of the most suitable raw article for factory processing is of the utmost importance to everyone concerned and the economic survival of the beet-sugar industry rests on this criterion alone. Apart from the skill of the grower in doing his share in the cheapest production of the best article possible, the success of the industry rests with suitable growing conditions and good varieties.

Any good soil can be used for growing sugar beet, and the best results cannot be expected on shallow, stony soils, heavy clays, and soils liable to suffer from drought in the summer, lateness in the spring, or undue wetness in the early winter when the time for lifting comes round. Sugar beet likes warmth and sunshine, and it grows best with summer temperatures of about 70° F., but it quickly suffers from drought and when temperatures are high over extended periods the sugar percentage is liable to be low. The actual accumulation of sugar in the roots takes place most satisfactorily during the cooler conditions of late summer and when there is ample sunshine for the plants to synthesise sugar after the root has reached its maximum size under the particular growing conditions. In this country suitable sugar beet climatic conditions are only found in the sunnier parts where there is a comparatively long growing season, and more often than not the effective growing season for the crop is either curtailed by late sowing or poor conditions

in the late summer and autumn. The fear of too high a proportion of the plants running up to flower ('bolting') is one of the
chief reasons why sugar beet is not sown as early in some seasons
as it might be, and this is one of the principal causes of a short
effective growing season for the crop.

Sugar beet, like mangolds, is an out-pollinated plant, and the
strains or varieties are not as pure breeding as is the case with
crops which are self-pollinated. There are undoubtedly different
races, even within one variety, which show adaptation to
particular growing conditions; some preferring cooler conditions,
others being more successful in drier climates than wetter ones,
and certain races showing suitability for the short growing
seasons with long days of the more northern latitudes. This form
of adaptation is important because it affects the maturity of the
root and the number of bolting plants, and selection by the plant
breeder of the most suitable types to grow has to take this into
consideration. Growing seasons vary considerably from year to
year in this country, as does the time of sowing according to the
ability of the grower to sow his crop under the particular soil
and weather conditions with which he has to cope, and the
same variety is exposed to a considerable range of growing
conditions in different areas and from season to season. There
are few crops in which varietal adaptation is as important as in
sugar beet because of the particularly unstable characters of the
plant and its susceptibility to changing climatic and weather
conditions.

Until very recent times sugar-beet seed was obtained exclusively from various countries in Europe, particularly
Germany, Holland, Sweden, Poland and Czecho-Slovakia.
With few exceptions, the seed was obtained from old-established
firms who bred strains for their own sugar-beet breeding
industries and sent stocks of their standard varieties to this
country. All the varieties cultivated in Britain were, therefore,
foreign bred and retained their continental names and type
descriptions until some firms established special breeding
stations and distributive organisations in this country. Within
the last ten years there have been efforts by English plant
breeders to breed varieties on a more extensive and intensive

scale, but this has meant for the most part merely re-selecting continental stocks under the conditions of this country with the object of isolating types more adapted to our sugar-beet growing conditions.

Varieties of sugar beet are classified into groups according to their characteristics with regard to root yield and sugar percentage. Those which have large roots are characterised by lower percentages of sugar and are designated as 'E' types, while the smaller rooted, but higher sugar percentage varieties are referred to as 'Z' types. The intermediate type with root size and sugar percentage between the 'E' and 'Z' types respectively are described as 'N' types. There are other types, but these three are the chief ones and the significant types from the agricultural point of view. The combination of root size and sugar percentage determines the theoretical amount of sugar being produced per acre of land, and on this criterion it is quite possible for varieties of different types to give the same theoretical sugar yields. The grower's object is to produce as great a weight of roots with as high a sugar percentage as possible under his particular growing conditions, and it is obvious that the number of roots to the acre as well as their size will determine the root yield, while root size and sugar percentage are determined by climate, soil, spacing and variety. The important matter for the grower is that the variety he chooses is suited to the growing conditions and the management of the crop, so that the fullest use is made of the particular combination of these three factors which will ultimately determine the cash value of his crop.

Other characters, besides root size and sugar percentage, which vary in different varieties are the conformation of the root including the tendency to fanginess, the juice purity; and the size of the foliage crown or 'top'. Root conformation is of importance both to the grower and the factories because it affects the cost of handling in the field; the weight of roots delivered to the factory after topping; the amount of soil ('dirt tare') removed at the factory; the efficiency of root processing, and the amount of sugar which can be extracted. The ideal conformation is a well-shaped and symmetrical root, broad across the 'shoulder' but free from greenness, buds and scars below the leaves, and with no fangs or coarse branches arising at any position along the length of the root. The leaf crown should be set neatly at the top of the root, which should grow almost entirely below soil level with a broad, clean taper to the tip. Such roots have least wastage when being topped in the field because the minimum amount of root is removed when cutting off the leaves, and freedom from fanginess means easier lifting and cleaning and less soil carted off the farm. Good conformation allows of more efficient treatment in the factory cutting machines, and weight for weight, a higher sugar content and extraction than from malformed roots.

In the extraction of the juice from the roots at the factory, soluble substances other than sugar are removed in solution. When these substances, particularly soluble nitrogenous compounds which comprise what is known as 'noxious nitrogen', occur in too high a concentration, the crystallisation of the sugar is more difficult, and the proportion recovered as refined sugar is less. The factory therefore requires as 'pure' a juice as possible to enable the highest possible extraction of the total sugar in the root.

The leaf crown and top of the sugar-beet root are not used in the factory and are removed by the farmer. While these tops are valuable as stock feed or for ploughing back as green manure, they may be an embarrassment when they grow too large in areas where they are not in much demand for either purpose. On very rich soils, such as good fens and silts, the tops are apt to grow excessively and they cannot readily be disposed of economically. On the other hand, on certain farms the tops are regarded as extremely valuable as stock feed, and may be used for folding sheep in the winter, feeding to cattle or making into silage. The nutritive value of the leaves and top of the root is equal to that of good swedes or some varieties of mangolds, and although this by-product in the main objective of producing sugar is an important consideration on many farms, it should not be allowed to divert the growers' attention from the primary function of the crop as a sugar producer.

Although all the foregoing characters are varietal, they are also much affected by the growing conditions. Early sowing usually increases the number of 'bolters', although some varieties can withstand the effects of early sowing better than others. Rich soils encourage large roots and 'tops'; shallow and stony soils tend to give a high degree of fanginess; weather conditions and fertiliser treatment affect root size, sugar percentage and juice purity; while bad attacks of diseases such as Virus Yellows and Downy Mildew can seriously reduce the all-round productiveness of the crop.

For general purposes over a wide range of soils, varieties of the 'E' type have been found to be the most remunerative and are the most extensively grown. The variety Kleinwanzleben E is the most popular and has given consistently good results over a number of years. More recently Cannell's 22 and 937, particularly the former, have come into use as varieties of similar performance to Kleinwanzleben E, while Johnson's Perfection (type E) which is derived from the Dutch Kuhn E and is very resistant to bolting, is a good variety with a root of rather higher sugar percentage and a somewhat smaller top. Garton's C also has a slightly higher sugar percentage than Kleinwanzleben E, but its roots tend to be longer. Where large-topped varieties are required, British S.K.W. or Goldsmith's Dobrovice may be used, but both show a greater tendency to bolt than the other varieties mentioned.

Of the 'N' type varieties suitable for cultivation in this country there are Johnson's Perfection (type N), Kuhn P, Garton's B and Webb's No. 2. There is not a great deal to choose between these varieties, and they all tend to be smaller topped and are resistant to bolting. The only 'Z' type variety grown to any extent in this country is Hillëshog, which is outstanding for its resistance to bolting and has medium small tops. Where lifting is planned to take place early, say before the middle of October, trials have indicated that it pays the grower to use a variety that has a high sugar percentage, and in all cases the length of the growing season, as dictated by the times of sowing and of harvest, is a matter for serious consideration in choosing the most appropriate variety.

Apart from general considerations of national economy and world sugar production, the future of sugar-beet growing on

British farms will be influenced strongly by the possibilities of breeding new and improved strains of beet by which sugar production in this country may be made more efficient. During the 1939–45 war a considerable home-grown sugar-beet seed industry was organised, and breeding investigations have been developed more extensively than at any previous time in the history of sugar-beet growing in Britain. In addition to the desirability of breeding strains which are capable of yielding more refined sugar for each acre of land devoted to the crop, there are important problems to be solved in relation to damage caused by diseases and pests. Considerable losses would be avoided if strains could be bred resistant to Virus Yellows, Downy Mildew and Beet Eelworm, and the means by which such improvements might be achieved are being studied.

A recent development in plant breeding which has been taking place principally on the European continent is the production of fodder beets by the hybridisation of sugar beets and mangolds. In some countries, notably Denmark, these 'sugar-mangolds' or 'half-sugar beets' are now being grown to the almost total exclusion of mangolds. The new hybrids are assessed essentially on their dry-matter production and palatability for livestock, and it is claimed that the best types are much superior to mangolds from the point of view of the most economic production of succulent fodder roots. A certain amount of work has already been done on this form of breeding in this country, but the results so far achieved have not proved as satisfactory as the claims made by the continental agriculturalists. This method of breeding fodder beets offers considerably greater scope for the production of forms with high dry-matter content than do methods which involve the use of mangolds only. Any improved fodder type, however, will have to possess the desirable root conformation of a good mangold, and also be resistant to bolting, if it is to become popular among farmers.

Chapter XVII

MISCELLANEOUS CROP PLANTS

There are certain old-established crops in this country which have never occupied large acreages, but some of which are relatively of a high cash value and of considerable economic importance locally to growers, as well as playing an important part in certain industries. The two most important examples are hops and flax, both of which are very interesting cultivated plants in their specialised production of commodities of considerable commercial significance. There are, in addition, crops such as buckwheat and linseed (which is botanically the same plant as flax), whose position in British agriculture is of minor importance except in times of war when they assume a greater economic value owing to the limitations of imports or price control on other crops. Of more recent introduction to the agriculture of this country are maize and sunflowers, the former already having won a very small but permanent position, and the latter still of a problematical nature as an economic crop.

From time to time small and localised areas of cultivation of specialised crops have been established in this country, some of which have been ephemeral and have been associated with a local industry, while others have maintained themselves on a very restricted basis. Tobacco, hemp, chicory, teasel and comfrey come into this category, while many others have been introduced but have failed to establish themselves as permanent features of our agriculture. It is as well to remember these successful and unsuccessful ventures because it is sometimes implied that agriculturalists are too conservative to try new possibilities and enthusiastic observers in other walks of life are apt to rail against the imagined waste of opportunity occasioned by the failure to strike out along new lines. It is fairly safe to assert that all the known cultivated plants which are likely to have any success in the agriculture of this country have from time to time been given their chance, and the natural resources in terms of climate and soil, together with the artificial opportunities in terms of economic markets, have been fully and thoroughly tested. At the moment it is most unlikely that there are any known cultivated plants of agricultural importance which could profitably be introduced for cultivation in this country, and the possibility is even more remote that any new crops of major agricultural significance are awaiting discovery. There is, of course, always the possibility that new methods in plant breeding may produce what is virtually a new crop plant, but this again is unlikely to be much more than an extreme variation of some well-established crop.

Of more immediate practical importance to the agriculture of this country is the gradual introduction of horticultural and vegetable crops as field crops. The case of the *Brassica* plants has already been mentioned, but there are in addition such vegetables as carrots, parsnips and onions, all of which may establish themselves as farmers' crops in the future. These plants have a steady and regular market and they grow well over a wide range of conditions in this country, and with the more important crops mentioned at the beginning of this chapter, are worthy of further consideration here.

Although hops are regarded as native to southern England many of the so-called wild hops found growing in the hedgerows are escapes from cultivation. The plant is an interesting one and belongs to the botanical order Urticales, which includes other useful cultivated plants of such diverse kind as the hemp, mulberry, fig, indiarubber fig and bread fruit. The hop (*Humulus lupulus*) is a member of the family Cannabinaceae which is one of six families in the Urticales, the other important economic one being the Moraceae, which, however, is mainly tropical in its distribution. One of the features characteristic of a large number of the plants in this order is that the flowers are unisexual, the two sexes being borne in some cases on the same plant, and in other cases on different plants. The hop is peculiar in having unisexual plants, and it is the female inflorescence that provides the valuable commercial product for which the plant is grown.

Hops were first cultivated extensively in this country in the sixteenth century, the principal sorts apparently having been introduced from Europe. The cultivation of this plant has always been centred in England and is now confined to eight counties—

Kent, Surrey, Sussex, Worcester, Shropshire, Essex, Hampshire and Berkshire—the whole area amounting to only some 20,000 acres of which Kent contributes by far the greatest proportion. The localisation of hop growing is due primarily to climate, but the highly specialised nature of the industry with its demands for labour, skilled in the handling of the crop, also contribute to the peculiar position of hop growing in this country. Being a perennial plant, hops occupy the same piece of ground for a number of years, the top growth, which has a twining habit, being trained to supports and dies back each year. The crop is propagated vegetatively by means of cuttings, and when a new area is being planted, setts are taken from established plants, it being usual to plant one male to 200 females.

The female inflorescences, called cones or burrs, are picked when ripe and dried on kilns. Their use in the manufacture of beer is due to the development of many small glandular hairs on the leafy bracts and perianths of the flowers. These glandular hairs secrete oils and resins which act as a flavouring and pre-servative for the beer. The full value of the cones is only developed after the flowers have been fertilised, and the flavour, aroma and preservative properties vary with the variety, the growing conditions, harvesting and drying. Varieties vary in their suitability for different soils and in their time of maturity and disease resistance. Four groups of hop varieties are recog-nised by the Hop Control Board—Goldings, Golding Varieties, Fuggle's and Tolhursts—each being characteristic of varieties possessing special characters and methods of utilisation. New varieties such as Fillpocket, Quality, Bullion and Brewer's Gold have been developed in recent years, and hybrids have been raised between the European *Humulus lupulus* and the American *Humulus americanus*, one of which has been put on the market under the name of Brewer's Favourite. These new varieties produce cones which are stronger in their commercial proper-ties, and it is therefore not necessary to use as much in brewing. Hops are very susceptible to various fungal and virus diseases, and in addition to the commercial characters—yield, ease of picking, vigour and time of ripening—disease resistance is an important varietal character.

Hops were formerly used medicinally, and it is interesting that in addition to their special functions of imparting flavour and aroma, acting as a preservative agent, precipitating proteins and performing a function in foam formation for the manufacture of beer, other special virtues are characteristic of this plant. The use of hops in brewing improves the dietetic properties of beer and stimulates digestion, as well as imparting special refreshing qualities to the beverage. Although many other plants such as tansy, wormwood, penny royal, mint and balsam have in the past been used in brewing, none combines such valuable characteristics as hops, which have now completely supplanted all other materials.

The plant known botanically as *Linum usitatissimum*, belonging to the family Linaceae, has a very long history as a cultivated crop for its fibre, which is extracted from the stem, and for its seed, which is one of the most valuable sources of oil in temperate regions. Although, botanically speaking, the same plant is used for both purposes, actually in practice two distinct types are recognised, and the crop is grown either for fibre or for seed. The fibre-producing types are taller growing and do not branch to the extent of the seed producers, and the former are grown in more northern latitudes of Europe, while the latter are Asiatic in origin and grow most successfully in warmer and more southerly areas.

Flax or linseed is an annual plant and the crop is taken in the rotation in various places according to the farming system and whether a fibre-producing or seed-producing variety is being grown. Flax for fibre is grown mostly in high rainfall areas, while linseed requires more sunshine and is confined mostly to the eastern parts of England. Of the 25,000 or so acres of both types cultivated in the United Kingdom in peacetime, by far the greatest proportion is flax for fibre, and most of this is grown in Northern Ireland. The crop is a very useful one to take on ploughed-up grassland because of its resistance to wireworm, but this is not a suitable practice on rich land because the plants become laid, and although heavy yields may be obtained, the percentage of fibre is usually low. Both linseed and flax do well after a cereal crop, and the range of soil types suitable for their cultivation is wide.

Fibre-producing plants have had a profound influence on the development of civilisation, and they come second only to food plants in this respect. Vegetable fibres have much more varied uses than animal fibres, and although many hundreds of fibre-producing plants are known and exploited, only a very few have any great commercial importance. Of these valuable plants the most important are very old, and flax ranks with cotton among the most prized fibre plants of the world. Flax is not grown so extensively as cotton, but it produces a higher-quality fibre which is used for the manufacture of linen, fine fabrics, writing material, threads and other commercial products.

The great value and particular uses of flax fibres are due to their special qualities for textiles and fabrics. High-quality fibres for these purposes are not common in plants, and they must possess special characteristics of a chemical and physical nature. They must be of sufficient length; fine, uniform in dimensions; durable and lustrous, with sufficient tensile strength, pliability and cohesiveness to be woven. Very few fibres possess all these attributes which make it possible to use them for making high-class fabrics; and flax fibres, which are pure cellulose in their composition, fulfil the requirements. It is important, also, from the economic and commercial points of view, that the fibres should be easily and cheaply isolated from the plant, and here again the flax plant offers little difficulty, the operations of retting, crumping and scutching, whereby the fibres are removed from the rest of the stem tissues, being simple and effective.

Considerable improvements have been effected in recent years in the varieties of flax available to growers in this country. For many years the stocks cultivated in the United Kingdom chiefly in Ireland—were of Russian and Dutch origin, and the first attempts at improvement were by mass selecting within these stocks. This resulted in the J.W.S. stocks which were an undoubted improvement on the originals, but these have subsequently given place to the Irish bred varieties, the best of which are Stormont Gossamer, Stormont Cirrus, Liral Monarch, Liral Crown, Liral Prince and Liral Dominion. These varieties have contributed materially to the efficiency of the flax industry by increasing the yield of fibre and the quality of the raw product for fine fabric manufacture.

Linseed is the most valuable source of drying oil and yields the most important oil of this type. The seed contains between 30 and 38 % of oil, but is also rich in protein. If the seed is to be used as a source of commercial oil production, it is crushed and the oil extracted, leaving behind a residue which is made into a meal or cake for cattle feed. Linseed cake may have up to 10% of oil, 30% of protein and 35% of carbohydrate and is a very valuable stock feed. Extracted cake is also sold which has only 3 % of oil and 36 % of protein, while the meal made from the whole seed has 36 % of oil. There is no other crop that can be grown economically in this country which possesses such high concentrations of oil and protein.

Several different stocks of linseed obtained from abroad used to be cultivated in this country, the best known being La Plata from the Argentine, Canadian, Russian and Bombay. Higher-yielding varieties better suited to this country are now available as the result of plant breeding in Canada. The most successful of those tried are Royal, Bison and Redwing, the first-named being superior to the other two.

Buckwheat is a plant of very minor importance in this country although considerable amounts are grown in Europe, Asia and North America. In spite of its name, it is not a true cereal and has no relationship with wheat or any of the Gramineae, the common English name merely being a corruption of the Dutch 'Bockweit' or the German 'Buch-weizen' meaning 'beech wheat' an allusion to the similarity of the fruit or 'grain' to a small beech nut. The plant belongs to the family Polygonaceae, of which we have no other cultivated representative in this country as a farm crop, although the rhubarb belongs to the same family. The generic name is *Fagopyrum*, and there are at least three species recognised—*esculentum*, *tartaricum* and *emarginatum*, the first-named being the species cultivated in this country. Buckwheat is probably a native of Central Asia and was taken into cultivation, as far as is known, about 2000 years ago, reaching China in the tenth or eleventh century and Europe in the Middle Ages.

The crop is a quick growing, succulent annual which is grown primarily for its 'grain', but may be used as a forage crop or green manure. It usually occupies no set place in the rotation and, owing to its short growing season, is commonly taken as a catch crop and may be sown later in the season than any other crop on the farm. It grows well on a wide range of soils, but its particular value lies in its ability to develop successfully on poor and acid soils which are unsuited to a true grain crop. It is remarkably free from insect and fungal attacks and can be grown on the same land year after year. Buckwheat is not suitable to wet cold soils or climates, and thrives best with plenty of sunshine and warmth. It is consequently found almost entirely in the eastern parts of England, although it is essentially a cool, temperate crop preferring relatively moist conditions.

The 'grain' has a comparatively thick and hard husk and is only fed in the whole condition to poultry or game. When cracked or ground it can be given in small quantities to stock and has a feeding value not unlike barley, except for the high fibre content. The 'grain' may also be milled, and the flour either used for making griddle cake and similar articles for human consumption, or else it may be fed to livestock; but it is deficient in protein because most of this is extracted in the middlings which form a useful feed for animals. As a green forage crop, buckwheat may be mixed with other plants such as peas and rape and the mixture used for cutting green and feeding to milking cows.

As a cultivated plant, buckwheat is an example of a relatively unimportant crop and the only representative in our agriculture of an insignificant botanical family, economically speaking, although there are a number of commonly occurring weeds, such as docks, sorrels, blackbindweed and knot grass, which are only too well known to farmers. Buckwheat is unlikely to attain a position of any real importance in British agriculture, and there has never been any justification for seriously attempting to improve the crop. There are, consequently, no very distinctive agricultural varieties from which it is important to select, and although types such as Japanese, Tartarian, Silver Hull and

Common Grey have all been cultivated in this country, only the last-named is of importance.

Maize, or Indian Corn, has been cultivated in this country on a small scale since it was originally brought to Europe in the sixteenth century from its native home in the New World. The plant, the botanical name of which is *Zea Mays*, is the only example of a major cereal belonging to the Gramineae which is a native of tropical and subtropical America. Although such a valuable crop in North and South America, where the great bulk of the total world production is grown, maize has never become important in this country or in the more temperate parts of Europe generally. The reason for this is that maize requires a high growing temperature, abundant sunshine, and a plentiful moisture supply, for its successful growth. It is extremely susceptible to frost, and cannot be sown safely before May in this country. To obtain anything like the maximum economic advantage of the crop it must grow quickly and set abundant seed which it can only do with a combination of high temperature, sunshine and moisture which is not commonly experienced over any considerable area of the British Isles. Low soil temperatures for germination and early growth, and lack of sustained conditions through the growing season suitable for rapid growth and seed setting and development, are the chief climatic limitations to the extended cultivation of this crop.

The primary use of maize is as a forage crop, the vegetative parts, immature grains and mature grains being used. As a forage, the crop can be cut in the green state and fed fresh or made into silage: in America and Canada maize is 'pastured off' with pigs, sheep or cattle when the cobs are ripening off and the leaves are still green. The ripe grain may be threshed and fed to stock in the crushed state or as meal, and is particularly valuable for its relatively high oil content. The grain may also be used for 'corn starch' manufacture, oil extraction, cereal foods and flour for human consumption, or for cooking in the whole state. As the basis of the livestock industry in North America, maize stands pre-eminent, and there are few crops that are put to so many uses when the less important 'sweet corn' and 'pop corn' types are also taken into consideration.

The difficulty about the full exploitation of the crop in this country is to find varieties which will set grain and reach an adequate stage of development for its best utilisation. There is a great range of types with regard to height of growth, tillering capacity and time of maturation, and the very late, robustly growing vegetative types such as the variety White Horsetooth, which never reaches the stage of grain production, can be grown successfully for cutting green and feeding to milk cows in the height of the summer when pastures are bare or suffering from drought. Earlier maturing types which will reach an adequate stage of grain development are required if full use of the crop is to be made, and in the southern and warmer parts of England this has, and can, be achieved in good seasons and with suitable varieties. The earliest types are mostly the shorter growing, higher tillering forms with the rounded or 'flint' grains, but there are also early varieties in the taller group with 'dent' grain which mature in reasonably good seasons. Varieties which have been used in this way in experimental work are Early Yellow Flint, Beacon, Gehu, Wisconsin 279 and Minnesota 13.

It is of the greatest importance that the right type and variety should be chosen in maize cultivation, and that the habits and characteristics of the available forms should be known by the grower. Lack of this knowledge and of experience have been responsible for a considerable amount of the failure in growing maize for various purposes in this country. Because the plant originally came from countries with hotter climates than this country it is commonly assumed that it is a great drought resister as an agricultural crop. This is not so: the most successful maize-growing areas in the U.S.A. are those which combine high summer temperatures with something like 16 in. of rain in the growing season, and it is entirely fallacious to assume that the crop can be grown economically in this country where summer droughts are associated with thin, dry soils.

Of the true vegetable crops which have assumed a greater importance in recent years as farm crops, those belonging to the botanical family Umbelliferae at present promise to be the most likely to retain a permanent economic position. The carrot (*Daucus carota*), the celery (*Apium graveolens*) and the parsnip

(*Pastinaca sativa*) are the most extensively grown, although the parsnip does not have anything like the importance of the other two and is still largely, or almost exclusively, a market-garden crop. Each of these plants occurs in the wild state in this country, and it is usually assumed that the cultivated forms have been derived from these wild species. The wild forms are annual or biennial plants in the case of the carrot and parsnip, and annual or perennial in the case of the celery, but the cultivated varieties are biennials.

The economic value of the carrot and parsnip depends on the edible and nutritious swollen tap root that is formed in the first year's growth of the plant. The food storage tissues in both cases are principally in the outer zone (cortex) of the root, but also in the central core of conductive tissue which, however, may become enlarged and woody and thereby reduce the feeding value. The chief food stored is sugar, and the carbohydrate content of carrots may be 9–10 %, and of parsnip over 11 %, while the carrot is also a valuable source of vitamin A, which is produced in the animal body from the carotene in the root.

The carrot is a considerably more variable cultivated plant with regard to root type and time of maturity than is the parsnip, and may be grown primarily for stock feed or as a table vegetable. The cattle types are slow growers and have long roots of white, yellow or red colour. The vegetable types, which are now the most extensively grown, have red roots, with a yellow or red central core, and the root shape varies from the long tapering tap root of the maincrop varieties to the short, 'stump-root' types which are early and may be used for forcing. All parsnips have long, white tap roots and require a longer growing season than the commonly grown carrots. The more restricted use and demand for parsnips have resulted in this crop never really passing out of the hands of market gardeners, whereas the greater popularity of the carrot, combined with its much longer season for marketing by using varieties which mature at different times, have resulted in a considerable expansion of carrot growing on the field scale, with the development of intensive cultivation of the crop on large acreages by specialist growers.

Such intensive carrot cultivation is confined to specially suitable carrot soils, which are light, deep and friable.

Celery is a crop of considerable commercial value on peaty, skirt and fen soils. It is a gross feeder, requires to be kept growing quickly, and is an expensive crop to grow. It is in the nature of a luxury vegetable and it is interesting in that it is the only farm crop which is grown for its expanded and succulent leaf stalks which form a 'heart' and compose the edible portion. There are white-stalked and pink-stalked types, the former being the more popular on the market, but the latter being the hardier. In the important celery-growing areas, the crop occupies a definite and major part in the rotation, and is an important cash crop on the farm.

Onions have sprung into prominence as a potential field crop largely as a consequence of the 1939–45 war. The plant belongs to the family Liliaceae, and its botanical name is *Allium cepa*. It is a biennial forming a bulb from the swollen bases of the leaf sheaths in which are stored sugar and other nutritive substances. There are many kinds and varieties of onions, some suited to autumn sowing, others to spring sowing, and some to forcing under glass. In addition, the colour of the skin of the bulb may be white, yellow, brown, red or purple, and the bulb shape flat, round, oval or elongated. While it is typical of a variety to have bulbs possessing predominantly some particular shape or colour, these characters vary in different stocks of the same variety, and even within the same stock there is some variability.

Varieties are grouped according to type and the characteristics of the original stocks, the more important being White and Brown Spanish, King Spanish, and Italian types including Rocca, Tripoli and the pickling Barletta and Silverskin forms. There are also the small shallot types, the Tree Onion, Potato Onion, Welsh Onion and Chives which are only garden or small-holding crops. There are many different varieties in the more important groups and the grower must first of all choose the type, then the variety, and finally obtain a good stock of that variety.

Although onions will grow satisfactorily on most types of soil, they prefer light to medium textures, but they will not withstand drought. Sunshine is necessary for ripening of the bulbs, and commercial onion growing is not practicable in areas of heavy late summer rainfall, because, although varieties vary considerably in their keeping qualities, it is not possible to obtain well-ripened bulbs, which are necessary for long storage, when the weather is wet. The crop may be grown on the same land year after year, and good growers spend considerable time and trouble on the preparation of their onion beds.

The vegetable crops which have been mentioned in this chapter cannot be regarded as typical crops of British agriculture, but they *are* typical of the type of undertakings which may develop in suitable areas of intensive arable farming if the trend towards the cultivation of perishable protective foods for human consumption should continue. There is a vast home market for such produce, and there is considerable scope for the small farmer situated in the right conditions of soil and climate, to take advantage of the demands for these various cash crops for direct human consumption.

It is characteristic of many of these vegetable crops that the difference between stocks of the same variety may be greater and of more economic importance than the difference between some of the varieties. The number of named varieties is in some cases very large indeed, and there is considerable confusion in varietal identity and nomenclature. Good and reliable stocks, of the best types and varieties for the growing conditions, methods of cultivation and markets are essential if these crops are to be grown to the best advantage and profit to the producer. Economic vegetable growing is also very much dependent on efficiently organised marketing which prevents undue loss and ensures that the product reaches the consumer in good condition.

The sunflower (*Helianthus annuus*) is an oil seed crop which has been grown economically on a field scale in other temperate countries, but has only been seriously considered for cultivation in this country during the last few years. The plant belongs to the very extensive family known as the Compositae, or 'Daisy family', and is valuable because the seed contains 30–35 % of oil and 17–20 % of protein calculated on the dry weight. The seed can be used for direct feeding to stock, or for extracting oil and using the residue to make a cattle cake which is not unlike linseed. It has been shown that under good conditions in England, yields comparable to those obtained in the important sunflower-growing countries can be obtained, but as yet the plant has not been taken into serious cultivation in this country. There are, however, indications that sunflowers might replace certain unreliable crops, such as the field bean, under some conditions.

Sunflowers appear to be adaptable to a wide range of soils, and the principal limitation to their cultivation in this country is climatic. The crop needs to grow quickly and to ripen while the conditions are sufficiently warm and dry in August or September. To ensure this, quick maturing varieties must be used, and of the varieties tested, Pole Star is the best in this respect, while Mars and Southern Cross, which are higher yielders, and somewhat later maturing, are good varieties for the drier parts of the country. The crop requires very careful handling in the field at harvest time to ensure adequate drying of the heads, but equal care is necessary to avoid loss through seed shedding when the heads are allowed to become too dry. At the moment the greatest difficulties to be overcome in the economic development of the crop are the avoidance of loss by birds during ripening, the devising of the most satisfactory methods of drying in the field and threshing the seed from the heads.

Owing to the lack of suitable plants in British agriculture for the production of concentrated edible oil and protein production, the economic cultivation of the sunflower would be a definite acquisition. The possible economic development of the crop in this country is dependent to a large extent on the future developments with regard to national economy and agricultural policy in this and other countries.

Chapter XVIII

SEED STOCKS AND IMPROVED
VARIETIES AND STRAINS

The highest standards of crop husbandry can only be maintained if the grower is provided with good seed, sound stocks and the best varieties possible. There are three means by which the grower's interests may be served, and each is indispensable in modern agricultural practice. First, it is necessary to see that there are ample supplies of seed of good quality and germination capacity. Secondly, there must be no shortage of sound and reliable stocks of the best varieties and strains available. Thirdly, attempts must be made to improve the individual crops by any plant-breeding methods that will rectify faults and improve the economic value of the crop by creating new varieties and strains. Each of these three means of maintaining crop husbandry at its most efficient level from the point of view of the standard of the stocks used, receives attention in this country, and a considerable organisation exists to provide the grower with the best material possible. This effort is wasted if it is not taken advantage of by the farmer, and there can be much loss of money and good food through failure to appreciate the importance of what is being done.

The Seeds Act of 1920, with its subsequent amendments, is the official recognition of the importance of providing the grower with the best seed sample possible. The object of the Act is to protect the farmer from unknowingly buying and sowing seed of inferior quality, and suitable means and legislation are provided to see that good seed is available. The Act requires that all seed coming within its scope should have been tested before sale to ensure as far as possible a satisfactory performance in the field, and a declaration has to be made to this effect. This declaration includes the percentage germination and purity, the presence of injurious weeds, and other specified information which is required for certain crops. In all cases it is necessary to state the kind of seed, and the name of the variety must be given for cereals and for red, white and crimson clover and sainfoin. For grass and clover seed, it is necessary to give the country of origin because of the extreme importance of this information for judging the value and behaviour of strains from different sources.

The advantages of such precautions and of the protection given to purchasers of seed are obvious, but it should be emphasised that the farmer is not forced to buy more expensive or better quality seed than he wishes to grow. But the loss which may result from sowing seed of poor germination capacity, or of the wrong strain or variety, and the danger of bringing on to the farm seeds of pernicious weeds which may be difficult to eradicate and which may contaminate the crop produce, are serious considerations. Good seed is not only the first requirement for successful field establishment of the crop and for high yields, but it also prevents one of the most potent causes of weed spread from operating. By ensuring that the desired variety or strain is purchased, the Seeds Act also prevents financial loss to the grower where it is important that the variety or strain should be adapted to the conditions, or suited to some particular form of utilisation or market.

A good seed sample of the most suitable variety or strain is then the first point for the purchaser to consider. Official Seed Testing Stations have been established for conducting the necessary tests, while many seeds firms possess their own facilities for doing this work. Although eye judgement of a seed sample can provide a useful guide to the value of that sample by showing such obviously undesirable features as poor size and filling, mechanical damage, bad harvesting conditions and poor storage, and the presence of obvious impurities such as weed seeds and dirt, this is not sufficient. A true assessment of the real value of a seed sample can only be obtained by properly conducted tests which are the only means of showing the germination capacity.

There is, however, one important character of a seed sample which neither the best tests nor the most critical eye can assess at all times with complete certainty. This is the purity of the stock and trueness to type according to the variety or strain that it purports to be. In some crops, such as wheat, barley and oats, it is possible to distinguish varieties by examining the grain, but even in these crops there may be some doubt whether a sample is or is not the variety whose name it bears, or whether there may not be impurities in the form of grain of another variety of very similar grain type. It is a simple matter to distinguish a white-grain impurity in a red-grain sample of wheat, for example, but it may be virtually impossible to state with certainty that a red-grain sample does not have a small proportion of another variety or varieties which have similar red grains. In other crops, such as swedes, turnips, sugar beet, mangolds, flax and many others, it is impossible to determine varietal purity from seed examination, although germination tests with observations on seedlings may show certain obvious inconsistencies or impurities. Similarly, it is impossible to identify strains of herbage plants by their seeds, although here again there is at least the practicability of distinguishing wild white clover from ordinary white clover, but no possibility of recognising certain local strains within each type.

It is then necessary to provide an additional means to the routine testing of seeds for germination and purity in order to ensure that the grower can purchase the kind of seed that he wants, by maintaining the identity of varieties and strains. Plant-breeding institutions, seeds houses, the National Institute of Agricultural Botany and local seed-growing associations all play their part in helping to maintain the 'trueness to type' and the varietal or strain purity of the individual crop plants. Crop inspection and certification of initial seed material is practised, while the plant breeder ensures a constant source of supply of authentic material by careful selection of original and nucleus stocks which can then be utilised for multiplication. The maintenance of good stocks, free from mechanical admixtures of other varieties and strains, is a comparatively simple operation in most crops once the dangers are understood, but there are other sources of stock deterioration and loss of purity which require more careful consideration.

The maintenance of the individuality of any stock is dependent in the first place on its reproducing itself in such a way that its important characteristics are not lost, and its type is maintained for as long as is desired. Some species of plants are inherently more stable than others in this respect, and in general a self-pollinated species is more constant, and it is easier to maintain cultivated stocks, than is the case with cross-pollinated species, unless the latter are vegetatively propagated as in the case of the potato. It is, for example, possible to select single plants which are for the most practical purposes 'true breeding' in self-pollinated crops, and to build up very uniform and constant stocks from such single plants. But even in self-pollinated species 'off-types' can arise, and in the case of hybrid varieties particularly, a constant watch has to be kept to ensure that the uniformity and character of the stock is being maintained. In cross-pollinated crops it is impossible to select single plants as initial material for building up stocks, but several individual plants have to be taken which may be used in various ways as the nucleus from which the stock is multiplied. Once such stocks are released for general cultivation they cannot be maintained for long on a commercial basis, and the plant breeder has constantly to supply new material for stock building if the individuality of the material is to be maintained.

Special certification schemes exist for particular crops as has been described in the chapters dealing with these crops. These certification schemes fulfil different functions in the various crops, but they all provide for inspection of the stocks and judgement as to their suitability for release to the grower. The most important of these schemes is that for seed potatoes, which provides for assessment of the virus content and relative freedom from bolters, wildings and other undesirable variations. The potato certification scheme is the greatest single factor in maintaining the standard of the stocks grown in this country, and is essential for maintaining the cultivation of the crop at an economic level. A different kind of certification scheme is in operation for wild white clover, the scheme merely operating to ensure that stocks which are to be marketed as true wild white comply with the necessary requirements with regard to their origin and are also of the desired type. In other herbage plants, provision is made by inspection for the certification of local strains and bred strains which must conform to the desired type before they are allowed to pass into distribution to growers.

During the 1939–45 war this country was in a difficult position with regard to imported seed, and there was a great stimulus to seed growing to make good the loss of foreign supplies. In addition to this temporary loss of certain classes of seed, there has been a great development in home-grown seed of herbage plants to meet the demands for indigenous and bred strains. This country has always grown all the seed required for some agricultural crops, and some of the seed of other crops, but it has depended on imported supplies for many kinds of seed for agriculture and horticulture. To meet the needs of the more intensive and diverse home-grown seed industry, a special Seed Production Committee works in conjunction with the National Institute of Agricultural Botany.

The climate of this country does not make seed growing easy, and seed production is mostly confined to particular areas where good harvesting conditions can be expected and the plants ripen their seed satisfactorily. The growing of many different kinds of seeds in a restricted area requires careful planning, co-ordination and supervision if the position is not to become chaotic, and special provision is necessary for handling large numbers of individual samples. Cross-pollinated crops particularly require careful attention to prevent contamination, and seed-growing areas for such crops as sugar beet and mangolds must be zoned, while the fields for individual strains of each crop must be sited to give reasonable safety from cross-pollination. Seed production is an exceedingly important part of the general organisation to maintain good standards of crop husbandry, and it requires carefully controlled attention and supervision. If there is a breakdown at this stage in meeting the grower's requirements, all the efforts in other directions are wasted.

While all these activities provide the required organisation for maintaining the existing stocks of established varieties and strains already available to the agriculturist, they are in themselves unable to provide for progressive improvement of crops through the production of new types. Crop improvement by plant breeding is the most important single means by which the level of production in terms of yield and quality can be systematically enhanced, apart, of course, from the activities of the grower in improving his methods of cultivation and the general level of his farming. The significant part played by plant breeding in the economy of British agriculture may best be judged by reference to the extent to which new varieties and strains of crops have been taken into cultivation during recent times, and since intensive efforts have been made in crop improvement. When the more important arable crops are examined on this basis, it will be seen that during the last fifty or sixty years there has been a constant changing of the varieties in general cultivation, and, with few exceptions, the old varieties have been replaced by improved varieties of more recent introduction. Even some of the old varieties which have survived the competition of the new ones, such as Squareheads Master wheat, have themselves been improved by selection of better stocks.

The work of crop improvement has been shared by private breeders, seeds merchants and state-aided institutions in this and other countries, and each can lay claim to a significant part in the production of new varieties for the British agriculturist. The cultivation of wheat, barley, oats, potatoes, sugar beet and various root crops, to mention only the more important arable crops of the country, is dominated by the comparatively recent introduction of new varieties, while the even more recent work on the improvement of herbage and forage plants is playing a most significant part in crop and stock husbandry. In the first spate of varietal improvement with the improved knowledge on breeding technique and the new consciousness of the possibilities, great improvements were effected and large numbers of new varieties appeared. As successive stages in improvement have been achieved, so has it become increasingly difficult to bring about further advances, and the number of new and improved varieties of outstanding merit reaching the grower is now smaller than it used to be, and the rate of improvement is necessarily slower.

The agriculturist tends to measure crop improvement in terms of increased gross yield, and it is sometimes argued that plant breeding has not achieved anything very considerable, because the average yield for the country to-day is very similar to what it was fifty or more years ago. Even if it is true that average yields have not improved as a result of the introduction of new varieties, it is not necessarily accurate to say that plant breeding has failed to improve the yielding capacity of crops. The statement is, in fact, not true, because it has been proved beyond all doubt that many of the new varieties of recent introduction are higher yielding than the old ones if given the appropriate and sufficiently fertile conditions to show it. But it is not generally understood that there is no one best variety that will out-yield all others under all conditions of soil, climate and farming, still less is it appreciated that a high-yielding variety bred for fertile conditions cannot be expected to be at its best on a poor, thin soil. Yielding capacity is a complicated character whose expression depends on the growing conditions, and the highest yielding variety for one environment is not necessarily the best for quite different conditions.

Although the ability to give high yields is an obviously desirable character in a new variety, yield must be considered in relation to the standard of farming, and the most important thing is to see that there are suitable varieties for different levels of farming and soil fertility. But absolute yield is by no means the only character of economic importance, and improvement can be achieved in other ways, the most important of which is quality. The idea of quality includes many attributes of a crop product, but it means essentially the suitability for a particular purpose. Baking and biscuit quality in wheat; malting, brewing and feeding quality in barley; feeding quality in oats; sugar percentage and juice quality in sugar beet; cooking quality and nutritive value in potatoes; fibre quality in flax; and feeding value as reckoned by the nutrient content in root crops, forage crops and herbage plants are all vitally important economic characters which are as important to consider as gross yields of the raw product. Indeed, it is no misstatement that yield is quite beside the point if the quality is so poor or badly adjusted to the utilisation of the product that it has no useful market. Two varieties of a particular crop may have similar yielding capacities under the same conditions, but the cash value of one may be infinitely greater than the other because of superior quality. It is in the improvement of quality that there is the greatest scope in breeding at the present time, because by this means the intrinsic value and financial returns from crop cultivation can be raised without having to adjust levels of farming and productivity.

Besides the basic improvement in crop yield and quality, plant breeding can, and has, contributed to the crop husbandry of this country by improvement in other characters or the development of new forms which enlarge the area of successful and economic cultivation of particular crops. Increased strength of straw in cereals has led to reduced costs of harvesting and prevention of damage to the grain, as well as making it possible to grow cereals successfully on rich soils. Shorter straw has eased the problems of combine harvesting, as also has the production of earlier maturing varieties. The development of winter-hardy varieties of oats and barley has relieved the strain on spring sowing and helped to 'stagger' harvesting, while in the case of oats, winter sowing reduces the risks of their cultivation in dry areas. In addition, both winter types and early maturing types makes it possible to cultivate these cereals in late districts because of their early maturation.

Crop improvement involves more than the production of new varieties or strains which are merely different from those already available to growers, and still more is it implicit that a new variety should be more than a new name. During the period of rapid multiplication of the number of varieties of crop plants in this country, many so-called new varieties were little more than old stocks renamed, or selections of old stocks with new names. The position became so serious in some crops that it was necessary to establish Synonym Committees for potatoes and cereals, the duties of which were to study and examine all new varieties and pronounce on the similarity to, or difference from, varieties already in existence. Where no recognisable differences exist, or where a new variety resembles an old one so closely as to make no practical significance to the grower, the new variety is said to be a 'synonym' of the old one.

The means adopted to ensure that growers have at their

disposal the best varieties, and that everything practicable is being done to improve the varieties available, involve considerable organisation of research and the straightforward study of crop plants, agricultural problems and industrial and consumers' demands. The simplest and cheapest method is to take advantage of improved varieties produced in other countries and import seed for trial in this country. This method is in constant operation, and plant breeders, seeds merchants and official crop-testing stations in this and other countries engage in the exchange of material in this way. Some of the most widely grown crop varieties at present in cultivation in this country have been obtained by this means, but for every variety that finds a place in agriculture, a large number either proves useless for direct practical use, or else is relegated to the varietal collections carried by plant-breeding institutes with a view to their possible use for breeding purposes.

The plant breeder himself can adopt several methods in his attempts at crop improvement, and his choice will depend on the particular crop with which he is working. The most obvious and simple method is by selection within good varieties already in cultivation, and a great deal of improvement has resulted in this way. The method is dependent for its success on the stock which is being subjected to selection either being mixed or not being pure breeding. In the early days of crop improvement all the standard varieties cultivated in this country were treated in this way, and improved stocks of some were produced. The mere act of 'purifying' the stock may be responsible for a distinct advance, particularly where quality is concerned, but it is obvious that there are strict limitations to this method of improvement, while in some circumstances it is possible to do harm by curtailing the range of conditions for successful cultivation. Sooner or later all the more valuable material has been utilised, and the pure stocks do not lend themselves to any further economic improvement by this means, except through the rare occurrence of a 'sport' or mutation which may offer itself, and the equally rare 'chance selection' of unknown origin.

The uses and possibilities of selection within material already available will depend on the kind of crop and the degree to which it has already been subjected to this method of improvement. Selection of this kind is the first thing to attempt when a crop is being initially studied, or when a new variety or stock is brought under observation. In self-pollinated crops the limits of any considerable improvement by this means are soon reached, and continued selection can only maintain the stock true to the type desired. In cross-pollinated crops, however, the power and scope of selection are greater because of the more mixed hereditary nature of the individuals composing a strain. Strict selection in cross-pollinated crops also becomes a matter of necessity to prevent deterioration, and whereas a self-pollinated stock may be multiplied and used indefinitely with a minimum amount of re-selection, in a cross-pollinated crop constant re-selection is required. In the case of herbage plants where native plants are used in agriculture, as in the grasses in this country, selection within the widely available material is a most valuable means of plant breeding, and is the basis of the development of indigenous and bred strains.

The improvement that can be effected by straightforward selection in existing stocks is limited by the amount and range of variation available, although, in cross-pollinated crops, improvement may result simply by selecting the best combinations of individuals or lines for crossing among themselves to exploit to the utmost any 'hybrid vigour' which they may show. But this exploitation of hybrid vigour is not strictly the result of pure selection, but is, as the term implies, the result of hybridisation, and it is to the technique of hybridisation that the breeder usually has to turn sooner or later in crop-plant breeding.

Hybridisation has two important functions in breeding—it provides the means of combining the desirable characters of two individuals in one individual, and it increases the range of variation for subsequent selection. To be most effective as a tool in plant breeding, hybridisation needs first of all to be carefully planned to fulfil some definite object, and secondly to be followed by adequate and efficient selection and testing. The actual operation of hybridisation is simply a technical manoeuvre, varying in the ease with which it may be effected according to the material being handled, which provides the basis of selection for the desired improvement. The success resulting from hybridisation depends primarily on selection—in the first place selection of the parents, and in the second place selection within the progeny over a number of generations. But there are definite limits to the powers of hybridisation in providing the desired combination of characters, even though the parents used possess the characters in question. For example, the hybridisation of an individual possessing high yield and poor quality, with another possessing low yield and high quality, does not necessarily mean that it will be possible to select a plant from the progeny in the second or subsequent generations which will combine the high yield of the one with the high quality of the other. The combination or recombination of desirable economic attributes is not a simple matter of mechanical reshuffling of parental characters. On the other hand, the possibilities offered not only by forming new combinations of characters, but also of more extreme expressions of individual characters, in one individual, make hybridisation the most valuable instrument for improvement possessed by the plant breeder.

The possibilities offered by hybridisation have been considerably enriched in recent years by improved technique which makes practicable the hybridisation of individuals belonging to different species, or even different genera. This, of course, does not apply to all species and genera, but to related groups of plants. The commonest and most widely practised form of hybridisation has always been between varieties of one species, in which there are no technical difficulties in making the hybridisation and in which there is no sterility of the immediate offspring. Such varietal hybridisation has provided the basis of plant breeding for many years, and has been the most important single means of crop improvement during the present century, but there has always been a definite barrier and limit to the 'wideness' of certain combinations once the breeder goes beyond, say, the varieties of one species. This barrier is due to the dissimilarity between the hereditary characters, and the chromosome structure and number, of the individuals which it is proposed to hybridise, resulting either in a complete failure to produce offspring or in varying degrees of sterility of the offspring if produced.

The method used to bring about these difficult hybridisations is essentially by doubling the number of chromosomes and thereby causing normal, fertile offspring to develop. This is most commonly done by using the drug colchicine which has most potent effects on the behaviour of the chromosomes in cell division, and by its application it has been possible to bring about parental combinations which have hitherto defeated all efforts. 'Wide-crossing' of distinct species and genera has opened up entirely new possibilities in the technique of plant improvement by hybridisation, and already the new creations are playing an important part in the breeding of some crop plants. It is now possible to visualise introducing the 'blood' and individual characters of plants possessing very valuable attributes, which previously had been thought to be quite beyond the practical technique of hybridisation.

Two other methods are open to the plant breeder as possible means of crop improvement, and both are concerned with 'artificial' manipulation of individual plants. The first of these involves doubling the chromosomes by the colchicine method previously mentioned, with the consequent production of 'polyploid' forms. Doubling the number of chromosomes sometimes produces a plant which differs from the original form in possessing larger individual organs or some special attribute of economic value expressed to a higher degree than plants with the normal chromosome number. It is now known that many of our species of cultivated plants possess multiple chromosome numbers, that is, they are polyploids, and artificial means of bringing about this condition are merely striving to achieve by controlled technical methods what has occurred naturally in many plants. The second of these 'artificial' methods is to try and increase the rate of production of sports or mutations by means of chemicals, X-rays, exposure to extreme temperatures and other technical refinements. Although it is undoubtedly possible to do this, it is not possible to control the kind of mutation, and the general experience has so far been that the new forms induced by the various treatments have little or no economic value as far as the improvement of crop plants is concerned. The method at the moment lacks the necessary degree of control which is essential for economic plant breeding.

The combined exertions of individuals, research institutes, seeds houses and other bodies concerned with the standard of seed stocks distributed to the farmer, are all directed to the principal objective of improving the efficiency of crop husbandry. It is only by a combination of good husbandry, good seed and good varieties that the best results are obtained, and no amount of skilled farming can make good the waste that comes from using anything but the really first-class stocks of the most suitable varieties, while bad farming wastes even the best of seed. The machinery exists for maintaining the standard of the stocks, and plant breeding is continually making its contribution of new varieties and strains, but it is the initiative of the grower that makes it possible to exploit to the fullest extent the available material.

INDEX